SpringerBriefs in Research and Innovation Governance

More information about this series at http://www.springer.com/series/13811

Fernando Ferri · Ned Dwyer
Saša Raicevich · Patrizia Grifoni
Husne Altiok · Hans Thor Andersen
Yiannis Laouris · Cecilia Silvestri

Governance and Sustainability of Responsible Research and Innovation Processes

Cases and Experiences

Springer

Fernando Ferri
National Research Council-IRPPS
Rome
Italy

Ned Dwyer
European Centre for Information on Marine Science and Technology
Lisbon
Portugal

Saša Raicevich
Italian National Institute for Environmental Protection and Research (ISPRA)
Chioggia
Italy

Patrizia Grifoni
National Research Council-IRPPS
Rome
Italy

Husne Altiok
Istanbul University
Istanbul
Turkey

Hans Thor Andersen
Aalborg University
Aalborg
Denmark

Yiannis Laouris
New Media Lab
Cyprus Neuroscience and Technology Institute
Nicosia
Cyprus

Cecilia Silvestri
Italian National Institute for Environmental Protection and Research (ISPRA)
Rome
Italy

ISSN 2452-0519 ISSN 2452-0527 (electronic)
SpringerBriefs in Research and Innovation Governance
ISBN 978-3-319-73104-9 ISBN 978-3-319-73105-6 (eBook)
https://doi.org/10.1007/978-3-319-73105-6

Library of Congress Control Number: 2017962021

Printed on acid-free paper

This Springer imprint is published by Springer Nature
The registered company is Springer International Publishing AG
The registered company address is: Gewerbestrasse 11, 6330 Cham, Switzerland

Preface

The scientific research is more and more addressing a holistic and multidisciplinary approach in the perspective of the Responsible Research and Innovation (RRI), aiming to anticipate and assess implications and societal expectations and to foster the design of inclusive and sustainable research and innovation. The concept of Responsible Research and Innovation is relatively recent and it is part of the Europe 2020 strategy with the objective to promote a vision for a stronger collaboration among social, natural, and physical scientists, societal actors and citizens in order to achieve a wider dimension of science and innovation, and improve the role of society in environmental preservation. Many researchers, European initiatives, and projects in different domains and contexts already started to address RRI principles and themes, but in the last years, a wide discussion is arising on these topics that requires to systematically share information and knowledge.

The **International Conference on Responsible Research and Innovation in Science, Innovation and Society (RRI-SIS2017)** is the first annual conference of the MARINA project[1] that aims at presenting case studies highlighting environmental preservation, RRI issues, principles, and topics related to the research and innovation in the context of the H2020 societal challenges. With this focus, the RRI-SIS conference was organized to cover and systematize these themes and to help fill the gap in the present literature. Several case studies and experiences on Responsible Research and Innovation performed in several past and running projects have been presented in this conference and collected in two books of the SpringerBriefs in Research and Innovation Governance. The first is titled "Governance and Sustainability of Responsible Research and Innovation Processes: Cases and Experiences" and the second "Responsible Research and Innovation actions in science education, gender and ethics: Cases and Experiences". The first book is given here, which builds on academic and industrial research and experiences that are being carried out at many different institutions around the world.

[1]The MARINA project has received funding from the European Union's Horizon 2020 Research and Innovation Programme under GA No. 710566.

We expect the book to serve as a valuable reference for a larger audience such as policymakers, decision-makers, researchers, and practitioners that seek to deepen their knowledge on practices, initiatives, applications, and experimental results of governance and sustainability in RRI processes. The 12 chapters of this book are organized into two parts, namely, Part I: Governance, Public engagement and Inclusion in the responsible R&D and Innovation process, and Part II: Sustainability of RRI processes.

Part I of the book includes seven chapters focused on experiences and case studies of policymakers and researchers to engage with the public and involve stakeholders in an early stage of policymaking. In Chap. 1, Forsberg et al. provide the first results of the RRI-Practice project aiming at identifying, understanding, disseminating, and promoting RRI implementation best practices at European and global levels. Chapter 2 examines the countermeasures to triple disasters (earthquake, tsunami, and NNP accidents) after the Fukushima accidents that Japanese government has taken in terms of governance, public engagement, and inclusion and innovation process. Chapter 3 describes the experiences including needs assessment, engagement, anticipation, reflection, and responsiveness in the transdisciplinary Assisted Living project. In Chap. 4, Ananasso et al. discuss resources and experiences, sharing medical records with the "crowd" to get ideas, perspectives, and cures. Chapter 5 presents some results from the ProGReSS Project investigating the dependency between innovative business models and the RRI framework in terms of inclusiveness and pro-poor growth. In Chap. 6, Flipse and Yaghmaei discuss a method to intuitively score performance on certain RRI dimensions and investigate its potential value as a strategy to implement RRI in industrial innovation practice. Chapter 7 describes some initiatives aiming at improving awareness of geographic information value through an inclusive process that creates awareness among citizens about the role they can have in decision-making processes.

Part II focuses on sustainability of RRI processes and includes five chapters. Chapter 8 presents the experience of a not-for-profit association, named Stati Generali dell'Innovazione, in matching the RRI approach with the EU strategy of promoting collaboration among societal actors (scientists, institutions, the political sector, citizens, etc.) in order to achieve a wider dimension of science and innovation. In Chap. 9, Magoni et al. test the RRI strategy in the aquaculture industry to improve fish quality by using the most suitable tools and strategies to reach several stakeholder classes. Chapter 10 discusses the socio-technical approach of the MARINA project and presents the results of the first series of Mobilisation and Mutual Learning workshops discussing RRI, societal and marine challenges. Chapter 11 presents a humanitarian laboratory aimed at bringing novel solutions by digging deep in refugees' contexts through applying the RRI process dimensions. Chapter 12 find out the evidence of responsible innovation theory in developing countries by considering a socio-technical system based on sustainability, the issue of risk, ownership, accountability, and responsibility.

The editors wish to acknowledge the following program committee of the RRI-SIS2017 Conference who rigorously review the chapters that were accepted for inclusion in this book:

Frederic Andres, National Institute of Informatics, Japan
Elena Aristodemou, CNTI, Cyprus
Alicia Betts, Global University Network for Innovation GUNi, Spain
Noemi Biancone, CNR, Italy
Chiara Bicchielli, CNR, Italy
Carmen - Gabriela Bostan, Institute of Educational Sciences, Romania
Maria Chiara Caschera, CNR, Italy
Chiara Cavallaro, CNR, Italy
Marta Cayetano, Universitat Autònoma de Barcelona, Spain
Alessia D'Andrea, CNR, Italy
Arianna D'Ulizia, CNR, Italy
Jozefien De Marrée, Vrije Universiteit Brussel, Belgium
Ander Errasti Lopez, GISME—University of Barcelona, Spain
Cristina Escrigas, Universitat Politècnica de Catalunya, Spain
Thomas Evensen, Research Council of Norway, Norway
Anna Maria Fleetwood, Swedish Research Council, Sweden
Paulina Forma, Universytet Jana Kochanowskiego, Poland
Ellen-Marie Forsberg, Oslo and Akershus University College, Norway
Tiago Garcia, Eurocean, Portugal
Agata Gurzawska, University Twente, The Netherlands
Tiziana Guzzo, CNR, Italy
Klaus Hadwiger, University of Hohenheim, Germany
Jesper Rohr Hansen, Denmark
Hideyuki Hirakawa, Osaka University, Japan
Manfred Horvat, Vienna University of Technology, Osterreich
Michela Insenga, University of Liverpool, UK
Adalheidur Jonsdottir, Icelandic Centre for Research—RANNIS, Iceland
Ewa Kocinska, Poznan Science and Technology Park, Poland
Georgios Kolliarakis, University of Frankfurt/Cluster of Excellence, Germany
Olga Kot, National Academy of Sciences of Ukraine, Ukraine
Martine Legris, Lille University CERAPS, France
Cèline Liret, Ocèanopolis, France
Allison Loconto, INRA, France
Adele Menniti, CNR, Italy
Jeanne Pia Mifsud Bonnici, University of Groningen, Netherlands
Pierpaolo Mincarone, CNR, Italy
Maura Misiti, CNR, Italy

Birgitta Myrman, Swedish Research Council, Sweden
Gabriele Nardone, Italian National Institute for Environmental Protection and Research (ISPRA), Italy
Francesco Niglia, University of Salento, Italy
Fabio Palazzo, CNR, Italy
Margus Pedaste, University of Tartu, Estonia
Marco Picone, Italian National Institute for Environmental Protection and Research (ISPRA), Italy
Francesca Ronchi, Italian National Institute for Environmental Protection and Research (ISPRA), Italy
Dalia Satkovskiene, Vilnius University/BASNET Forumas association, Lithuania
Kjersti Sjaatil, Oslo and Akershus University College of Applied Sciences, Norway
Astrid Souren, EARMA/Radboud University, Netherlands
Veronika Tamas, Hungarian Academy of Sciences, Centre for Social Sciences, Hungary
Ruud ter Meulen, Centre for Ethics, University of Bristol, UK
Asta Valackiene, Kaunas University of Technology, Lithuania
Adriana Valente, CNR, Italy
John Walls, University of Glasgow, Scotland
Miyoko Watanabe, JST, Japan
Kirstie Wild, Danish Technological Institute, Denmark
Go Yoshizawa, Osaka University, Japan

Moreover, the editors wish to acknowledge the following chairs and committees of the RRI-SIS2017 Conference for their support to the organization of the conference.

Advisory Chairs

Giuseppe Borsalino, European Commission, Belgium
Zakaria Benameur, Research Executive Agency, European Commission, Belgium
Alex Bielak, Alex Bielak Communications, Canada
Ana Isabel da Silva Araújo Simões, Universidade de Lisboa, Portugal
Maria Paula Diogo, Universidade Nova de Lisboa, Portugal
Adrian Slob, TNO, Netherlands
Nina Zugic, UK

Science Communication Chairs

Iwona Gin, Nausicaa, France
Liina Vaher, AHHAA, Estonia
Manuel Cira, ROM WON, Belgium

Publicity Chairs

Elena Giusta, Italian National Institute for Environmental Protection and Research (ISPRA), Italy
Mara Gualandi, APRE, Italy
Emanuela Danè, APRE, Italy
Micaela Candea, MareNostrum, Romania

Industrial Chairs

Xenia Schneider, XPRO, Cyprus
Mato Knez, NanoGUNE, Spain
Eoin Nicholson, SmartBay, Ireland

Local Organizing Committee

Noemi Biancone, CNR, Italy
Chiara Bicchielli, CNR, Italy
Maria Chiara Caschera, CNR, Italy
Alessia D'Andrea, CNR, Italy
Arianna D'Ulizia, CNR, Italy
Tiziana Guzzo, CNR, Italy
Gabriele Nardone, Italian National Institute for Environmental Protection and Research (ISPRA), Italy
Marco Picone, Italian National Institute for Environmental Protection and Research (ISPRA), Italy
Francesca Ronchi, Italian National Institute for Environmental Protection and Research (ISPRA), Italy

The editors wish to acknowledge the organizing committee of the EARRI Workshop that has contributed to the success of the RRI-SIS conference:

EARRI Organizing Committee

Angelo Corallo, University of Salento, Italy
Francesco Niglia, University of Salento, Italy
Carlo Maria Medaglia, Link Campus University, Italy
Valentina Volpi, Link Campus University, Italy
Peggy Valcke, University of Leuven
Natalie Bertels, University of Leuven

Finally, the editors wish to acknowledge the secretary of the RRI-SIS conference for its valuable support to the organization of the conference:

Eugenia Bellocco, CNR, Italy
Cristiana Crescimbene, CNR, Italy
Gianni Galli, CNR, Italy
Angelo Perugini, CNR, Italy
Laura Sperandio, CNR, Italy
Wanda Toffoletti, CNR, Italy

Rome, Italy	Fernando Ferri
Lisbon, Portugal	Ned Dwyer
Chioggia, Italy	Saša Raicevich
Rome, Italy	Patrizia Grifoni
Istanbul, Turkey	Husne Altiok
Aalborg, Denmark	Hans Thor Andersen
Nicosia, Cyprus	Yiannis Laouris
Rome, Italy	Cecilia Silvestri

The original version of the book was revised: For detailed information please see erratum. The erratum to the book is available at https://doi.org/10.1007/978-3-319-73105-6_13

Contents

Contributors

Husne Altiok İstanbul Üniversitesi Merkez Kampüsü, İstanbul, Turkey

Fulvio Ananasso Stati Generali dell'Innovazione and CDTI, Rome, Italy

Hans Thor Andersen AAU, Aalborg, Denmark

Elena Aristodemou CNTI, Nicosia, Cyprus

Margot Bezzi APRE, Rome, Italy

Noemi Biancone IRPPS—CNR, Rome, Italy

Chiara Bicchielli IRPPS—CNR, Rome, Italy

Mathis Bogens Science Centre AHHAA Foundation, Tartu, Estonia

Chiara Bolognini Italian National Institute for Environmental Protection and Research (ISPRA), Rome, Italy

Chiara Buongiovanni APRE, Rome, Italy

Antonia Bierwirth Tecnalia Reseach and Innovation, Donostia/San Sebastián, Spain

Stefano Bertacchi University of Milano-Bicocca, Milan, Italy

Paola Branduardi University of Milano-Bicocca, Milan, Italy

Ausilia Campanaro University of Milano-Bicocca, Milan, Italy

Mihaela Mirea Candea Mare Nostrum, Constanţa, Romania

Maria Chiara Caschera IRPPS—CNR, Rome, Italy

Manuel Cira ROM-WON, Brussels, Belgium

Emanuela Danè APRE, Rome, Italy

Ned Dwyer EurOcean, Lisboa, Portugal

Alessia D'Andrea IRPPS—CNR, Rome, Italy

Arianna D'Ulizia IRPPS—CNR, Rome, Italy

Dilek Ediger İstanbul Üniversitesi Merkez Kampüsü, İstanbul, Turkey

Sergio Farruggia Stati Generali dell'Innovazione and AMFM GIS Italia, Genoa, Italy

Fernando Ferri IRPPS—CNR, Rome, Italy

Steven M. Flipse Delft University of Technology, Delft, Netherlands

Ellen-Marie Forsberg (HiOA) Work Research Institute, Oslo and Akershus University College, Oslo, Norway

Yuko Fujigaki The University of Tokyo, Tokyo, Japan

Andrea Galimberti University of Milano-Bicocca, Milan, Italy

Tiago Garcia EurOcean, Lisboa, Portugal

Anca Gheorghe Mare Nostrum, Constanţa, Romania

Iwona Gin Nausicaà, Boulogne-sur-Mer, France

Elena Giusta Italian National Institute for Environmental Protection and Research (ISPRA), Rome, Italy

Nagore Ibarra Gonzalez CIC nanoGUNE, Donostia–San Sebastian, Spain

Patrizia Grifoni IRPPS—CNR, Rome, Italy

Mara Gualandi APRE, Rome, Italy

Raúl Tabarés Gutiérrez Tecnalia Reseach and Innovation, Donostia/San Sebastián, Spain

Tiziana Guzzo IRPPS—CNR, Rome, Italy

Helin Haga Science Centre AHHAA Foundation, Tartu, Estonia

Jesper Rohr Hansen AAU, Aalborg, Denmark

Amira Buz Khallouf Syrian Social Innovators, Lisbon, Portugal

Mato Knez CIC nanoGUNE, Donostia–San Sebastian, Spain

Massimo Labra University of Milano-Bicocca, Milan, Italy

Miltos Ladikas Institute for Technology Assessment and Systems Analysis (ITAS), Karlsruhe Institute of Technology (KIT), Karlsruhe, Germany

Yiannis Laouris CNTI, Nicosia, Cyprus

Chiara Magoni University of Milano-Bicocca, Milan, Italy

Shilpa Mishra School of Social Sciences, Centre For Studies in Science Policy, Jawaharlal Nehru University, New Delhi, India

Eoin Nicholson SmartBay Ireland, Rinville, Oranmore, Galway, Ireland

Richard Owen University of Exeter Business School, Exeter, UK

Angelica Paiu Mare Nostrum, Constanţa, Romania

Marco Picone Italian National Institute for Environmental Protection and Research (ISPRA), Rome, Italy

Chiara Pesciaroli University of Milano-Bicocca, Milan, Italy

Renzo Provedel FareImpresa, Genoa, Italy; Stati Generali dell'Innovazione, Rome, Italy

Saša Raicevich Italian National Institute for Environmental Protection and Research (ISPRA), Chioggia, Italy

Francesca Ronchi Italian National Institute for Environmental Protection and Research (ISPRA), Chioggia, Italy

Xenia Schneider XPro Consulting, Strovolos, Nicosia, Cyprus

Monica Sebillo AMFM GIS Italia, Stati Generali dell'Innovazione, University of Salerno, Fisciano, Italy

Cecilia Silvestri Italian National Institute for Environmental Protection and Research (ISPRA), Rome, Italy

Rajbeer Singh School of Social Sciences, Centre For Studies in Science Policy, Jawaharlal Nehru University, New Delhi, India

Clare Shelley-Egan (HiOA) Work Research Institute, Oslo and Akershus University College, Oslo, Norway

Erik Thorstensen Work Research Institute, Oslo and Akershus University College (HiOA), Oslo, Norway

Emad Yaghmaei Delft University of Technology, Delft, Netherlands

Part I
Governance, Public Engagement and Inclusion in the Responsible R&D and Innovation Process

Chapter 1
Implementing Responsible Research and Innovation in Research Funding and Research Conducting Organisations—What Have We Learned so Far?

Ellen-Marie Forsberg, Clare Shelley-Egan, Miltos Ladikas and Richard Owen

Abstract Responsible research and innovation (RRI) has emerged in recent years, especially in Europe, as a science policy framework that (a) seeks to align technological innovation with broader social values and (b) supports institutional decisions concerning the goals and trajectories of research and innovation under conditions of uncertainty, ambiguity and ignorance. Rather than relying simply on consumer choice and market mechanisms on the one hand, or risk-based regulation on the other, RRI seeks to determine what constitute the goals, purposes and trajectories of (and alternatives to) technoscience and innovation, and thus the directions towards which these should be oriented, suggesting that these should be underpinned by shared public values. In addition to this overall philosophy of RRI, the European Commission has focused on five constituent policy keys (sometimes called pillars) of RRI that have their historical roots in the Science-in-Society programme; namely societal engagement, gender in research, open access, science education, and ethics. Action on these keys is seen as integral to an RRI approach and to Europe's ability to respond to societal challenges. A further issue in the European context concerns how to 'federate' the RRI community in the EU and promote institutional changes to foster RRI in research

The original version of this chapter was revised: For detailed information please see Erratum. The erratum to this chapter is available at https://doi.org/10.1007/978-3-319-73105-6_13

E.-M. Forsberg (✉) · C. Shelley-Egan
Work Research Institute, Oslo and Akershus University College, Oslo, Norway
e-mail: ellenmarie.forsberg@hioa.no

C. Shelley-Egan
Akershus University College (HiOA), Pilestredet 35, 0166 Oslo, Norway

M. Ladikas
Institute for Technology Assessment and Systems Analysis (ITAS),
Karlsruhe Institute of Technology (KIT), Karlsruhe, Germany

R. Owen
University of Exeter Business School, Exeter, UK

F. Ferri et al., *Governance and Sustainability of Responsible Research and Innovation Processes*, SpringerBriefs in Research and Innovation Governance,
https://doi.org/10.1007/978-3-319-73105-6_1

institutions (a topic addressed at the European Commission RRI conference in Rome November 2014). This implies engaging stakeholders, research organisations, universities, funding agencies and public authorities in RRI. Some European research conducting and research funding organisations have begun to make formal policy commitments to RRI; others have developed RRI programmes and others still have embedded explicit RRI elements within broader programmes of emerging technologies and innovation. The European Commission's 'open to the world' agenda implies involving non-European countries in the RRI discourse. However, beyond Europe, in emerging economies in the Global South (Brazil, India and China) and also in some advanced economies (Japan, Australia), there is little awareness of the concept of RRI, although some elements of the EC's constituent keys have been taken up as thematic priorities by national research organisations. Considerable work needs to be done before RRI is recognised as a concept that offers traction in non-European contexts and research initiatives. There is a dearth of research that has assessed the challenges, efficacy and impact of the ongoing programmes on RRI, partly due to a lack of standardised methodologies that would be required to produce comparative results, and partly because these initiatives are themselves quite new. The project Responsible Research and Innovation in Practice (RRI-Practice), funded by the European Commission Horizon 2020 Science-with-and-for-Society programme (grant no 709 637), is an attempt to respond to this situation. The RRI-Practice project intends to advance European and global awareness of RRI, support its implementation in practice and provide a solid empirical knowledge base on RRI implementation. The main aim of RRI-Practice is to analyse RRI related discourses and pathways to implementation, including barriers and drivers, in a number of research conducting and research funding organisations worldwide, in order to identify, understand, disseminate and promote RRI implementation best practices that can be scaled up at European and global levels. The project started September 2016 and has so far concentrated on mapping the national RRI discourse in the 12 partner countries. As part of this work, national workshops have been held. This paper will present the analytic concept of the project and the results from the workshops, and will reflect on challenges identified in the work so far.

1.1 Background—on RRI

Responsible research and innovation (RRI) has emerged in recent years, especially in Europe, as a science policy framework that (a) seeks to align technological innovation with broader social values and (b) supports institutional decisions concerning the goals and trajectories of research and innovation under conditions of uncertainty, ambiguity and ignorance. Rather than relying simply on consumer choice and market mechanisms on the one hand, or risk-based regulation on the other, RRI seeks to determine what constitutes the goals, purposes and trajectories of (and alternatives to) technoscience and innovation suggesting that these should be underpinned by shared public values.

In addition to this overall philosophy of RRI, the European Commission (EC) has focused on five constituent policy keys of RRI that have their historical roots in the Science-in-Society programme; namely societal engagement, gender in research, open access, science education, and ethics.[1] Action on these keys is seen by the EC as integral to an RRI approach and to Europe's ability to respond to societal challenges.

A further issue in the European context concerns how to 'federate' the RRI community in the EU and 'mainstream' institutional changes to foster RRI in research institutions. This implies engaging stakeholders, research organisations, universities, funding agencies and public authorities in RRI. Some European research conducting and research funding organisations have begun to make formal policy commitments to RRI; others have developed RRI programmes; and others still have embedded explicit RRI elements within broader programmes, especially in the field of emerging technologies and innovation. Moreover, most organisations commit at least to some of the EC RRI policy keys, although they usually do not frame this commitment explicitly under the banner of RRI.

The EC's recent 'open to the world' agenda implies also involving non-European countries in the RRI discourse. However, beyond Europe, in emerging economies in the Global South (Brazil, India and China) and also in some advanced economies (Japan, Australia), there is little awareness of the concept of RRI, although some elements of the EC's constituent keys have been taken up as thematic priorities by national research organisations (see Brom et al. 2015). Considerable work needs to be done before RRI is recognised as a concept that offers traction in non-European contexts and research initiatives. By engaging with major global S&T players and their sometimes different and often differentiated needs, it may become clear that research and innovation may have to be 'responsible' in ways that are not an immediate priority for those nations where the RRI discourse has so far largely developed (Macnaghten et al. 2014).

1.2 Implementing RRI

RRI discourses have, in a Foucauldian sense, become constitutive in nature in that they can result in tangible activities, behaviours and outcomes. There have been experiments in implementation, which in turn have informed its theoretical conceptualisation (see e.g. Owen and Goldberg 2010; Macnaghten and Owen 2011). In addition, many research organisations and funding organisations have long established practices for promoting open access, ethical standards, or gender equality, items that could now be classified under the umbrella of RRI. We know from these experiments and practices that attempt to implement RRI across regional, national and local contexts will require a firm understanding of its framing and of the context

[1]In some policy documents 'governance' is noted as a sixth key.

of implementation, including cultural, political and institutional dimensions (Jacob 2013). However, there is a dearth of research that has assessed the challenges, efficacy and impact of the ongoing programmes on RRI, partly due to a lack of standardised methodologies that would be required to produce comparative results, and partly because these initiatives are themselves quite new.

1.3 Studying RRI Practices

The project Responsible Research and Innovation in Practice (RRI-Practice), funded by the European Commission Horizon 2020 Science-with-and-for-Society programme (grant no 709 637), is an attempt to respond to this situation. The RRI-Practice project intends to advance European and global awareness of RRI, support its implementation in practice and provide a solid empirical knowledge base on RRI implementation. The main aim of RRI-Practice is to analyse RRI related discourses and pathways to implementation, including barriers and drivers, in 24 research conducting and research funding organisations, in 12 European and non-European countries, in order to identify, understand, disseminate and promote RRI implementation best practices that can be scaled up at European and global levels.

The project started September 2016 and has so far concentrated on mapping the national RRI discourse in the partner countries. As part of this work, national workshops in twelve countries across the globe have been held. In the remainder of this paper we will present the research approach and the analytic concept of the project and reflect on findings from the workshops related to the analytical concept.

When studying the implementation of RRI in research organisations at both European and global levels it must be noted that there is still considerable debate in Europe as to relevant processes involved in RRI approaches. Moreover, the concept itself has also yet to enter mainstream S&T debates in non-European countries. A realistic discussion on possible implementation strategies and assessment processes has to therefore investigate and understand not only the RRI keys, as they are defined by the EC, but also the organisations' own framing of responsibility in research and innovation (what might be described as 'de facto RRI'). An example of this is the UK Engineering and Physical Sciences Research Council's (EPSRC) policy for responsible innovation as 'anticipate, reflect, engage and act'. It is therefore important to take a reflective, learning approach to such implementation rather than a top-down paternalist approach.

We also suggest that RRI should be considered a *process* that requires reflection on its *normative orientation* and the institutional values on which RRI activities are anchored. At EC level it has been suggested that normative anchor points for RRI are defined within the Lisbon Treaty (von Schomberg 2013), but these will inevitably vary across countries and specific organisations according to their own mission statements and priorities. This then requires sensitivity towards the organisations' own values, as they are embedded in different national and organisational contexts. More generally, there cannot be a fixed understanding of RRI 'implementation', but rather a family resemblance (Wittgenstein 1953) with common denominators in a plurality of practices that need to be taken as a starting point instead of an assumed essence of the concept.

Furthermore, one needs to be realistic as to the timeframe that RRI implementation would require in order to allow for proper assessment of its processes and impact, especially when one considers the potential for implementation at a global level. Programmatic changes at organisational levels require years of preparation, followed by years of execution, in addition to a considerable amount of time for assessment. Implementation programs on RRI can both challenge and build on *existing policies and practices* that are already embedded in organisations, using the expertise and commitment of local stakeholders to better understand and, where necessary, adapt such practices towards RRI goals. This does not imply neglecting the role of mutual learning and transfer of best practices and lessons learned *between* organisations.

1.4 The Importance of Understanding Organisations

Organisational practices with regard to, for example, gender equality, open access or anticipation are influenced by a diverse set of contextual and internal forces. Some of these are institutional or national dynamics established and maintained beyond the scope of single organisations; such as legislative prerequisites (e.g. gender quotas), established peer review practices, academic incentive systems, patenting and commercialisation reward systems or the (in)transparency of overall policy making. Some other dynamics relate to organisational particularities, often developed over a significant amount of time, and embedded in distinct organisational identities (e.g. programmatic excellence). Change can also be the result of individuals' formal or informal agenda setting or implementation power (i.e. change champions). Even only one initial RRI promoter in an organisation may create long-term organisational change (Hardy and Maguire 2008). RRI leadership is, therefore, another important factor for encouraging broader uptake of RRI. In total, the norms, roles and governance arrangements of the societal and institutional context and the RRI practices of specific organisations and individuals are inter-related.

Therefore, in trying to increase the uptake of RRI, one needs to take into account the incentives and motivations of actors, the institutional contexts in which RRI practices develop and that promote or constrain the behaviours of actors, and consider the mission and manoeuvring space of the organisations and individuals that own the practices.

Institutional theory is widely used to analyse public administration, and it has, for example, been used to study governance (Cashore 2002; Forsberg 2012a), change management (Henisz and Zelmer 2003), organisational ethics (Boyle et al. 2001; Forsberg et al. 2012) and innovation systems (Moodysson and Zukauskaite

2012; Forsberg 2012b). Scott (1987) distinguishes between studying institutions as 'rational', 'natural' and 'open' systems. This has been interpreted in an ethical context by Boyle et al. (2001), who claim that many of the ethical challenges of organisations can be understood only by taking these different dimensions into account. This approach is also useful for understanding organisational dynamics related to implementing RRI.

According to Boyle et al. the rational systems approach understands the organisation as having '(1) a visible set of hierarchical authority relations in which (2) work activities are governed by formal rules and clearly defined criteria for evaluation, relations that (3) are designed to pursue some set of goals' (2001, p. 31). From an RRI perspective, the rational system includes formal mandates, hierarchies, ethical guidelines and codes of conduct, monitoring and assessment systems, etc. and studying this also involves scrutiny of workload, the availability of resources, etc. (see also Forsberg et al. 2012). Understanding the organisation as a *natural* system involves scrutinising the informal sides of the organisation. This allows the study of how individuals may have a significant influence on attitudes and conduct in an organisation and how cultures and sub-cultures may flourish more or less decoupled from formal structures, often undermining, replacing or transforming them (Boyle et al. 2001, p. 33). Boyle et al. argue that in an ethical context (and, we can add, an RRI context) this means to 'identify the degree to which actual norms, rules, and practices differ from official or formal ones, and then to enquire into how the informal counterparts influence production outcomes, interpersonal relations, and goal attainment' (ibid., p. 33). Boyle et al. state that although there is a need for both formal and informal systems in an organisation, a gap between them that is too wide might engender hypocrisy, cynicism and disillusionment.

Finally, an *open* systems approach to organisations focuses on the relationships between the organisation and its environment, on which it is 'dependent for resources, personnel, and legitimacy' (ibid., p. 35). For instance, for research conducting and research funding organisations, national Ministries of Research and Innovation represent key contextual agents.

In a similar way as Forsberg et al. (2012) have shown in the case of designing effective ethics programs in a military organisation, studying and implementing RRI in research organisations needs to take into account the dynamics between the rational, natural and open systems and understand how these imply barriers and drivers for the RRI agenda. A potential barrier for an RRI key might, for instance, be that even if gender balance is described as a central value in official documents, leaders communicate that this is of little importance compared to the value of research excellence. Or, conversely, individuals in an organisation may succeed in influencing co-workers to start an open access journal even if there is no formal decision made in their organisation about such a strategy.

We can summarise the above approach to organisations studies in Table 1.

As it seems useful to clarify challenges and conditions for successful RRI work, we will in the RRI-Practice project use this analytic framework to develop organisational RRI reviews and Outlooks.

Table 1 Framework for studying the included organisations

	Rational system	Natural system	Open system
Aspects of organisations	Mandates, legislative frameworks, formal hierarchies	Culture, informal routines, informal reward systems, focus on management	Policy learning, pressures from key stakeholders (owners, the public, etc.)
Potential drivers for RRI	Active ownership (e.g. the state), legislation that includes social responsibility as a core element of the mandate, formal evaluation criteria adapted to RRI goals	RRI dimensions become mainstreamed, managers start seeing RRI dimensions as an obvious part of their responsibilities, no social acceptance for neglect of the RRI dimensions	Pressure from the media, success stories from organisations considered to set 'gold standards' in the field
Potential barriers to RRI	No formalised pressures to conform to RRI dimensions	Informal incentive systems reward economic output/ excellence/etc., effectively marginalising the RRI dimensions	Important stakeholders reward, for instance, excellence and economic performance to a greater extent than RRI related matters
Methods	Analysis of formal documents	Interviews with employees at different levels in the organisations, focus groups	Media analysis, interviews with top management

1.5 Findings so Far

So far the project has focused on mapping the understanding of responsibility in research and innovation in the different partner countries and among different stakeholders, as well as mapping relevant practices. This is important, especially for studying the organisations as open systems, which operate in a context where meanings are translated and expectations negotiated. RRI workshops have been held in all twelve partner countries. The workshops have shown a surprising willingness of stakeholders in all countries to discuss their policies and practices in RRI terms,—even when they were not initially familiar with the concept.

Stakeholders that were unfamiliar with the concept of RRI focused mostly on the EC RRI keys since they could readily identify national debates and ongoing activities related to RRI framed as ethics, gender equality, public engagement and open access. Science education, however, came across as a vague concept that requires a more detailed description and was hence generally less addressed. Some appreciate RRI as an umbrella concept that could integrate policies on the different aspects that were presented as keys. Those organisations that were familiar with the notion of RRI, were more sceptical to reducing it to a focus on keys. These organisations already had policies and activities on the RRI keys, but considered RRI to indicate a deeper systemic change in the relation between science and society. This indicates that the keys might be positioned as preconditions or prerequisites for a broader imaginary of a responsible innovation system involving reconfiguration of knowledge flows and institutions. In this respect perhaps RRI could be seen as a process of development. If this is the case, the weight put on strengthening the EC keys versus e.g. the AIRR dimensions (anticipation, inclusion, reflexion and responsiveness) may need to be adapted to different national contexts depending on where they are on this development journey.

When regarding RRI as a commitment to anchoring research and innovation better in societal needs, another topic also becomes important, namely who is to define such needs. Does it imply more state steering of research and innovation or does it mean more public engagement in science? And what is the balance between deliberation and representation? The cultural context of the research organisations in different countries has great impact on how the RRI keys (like public engagement) are regarded, and to what extent they are emphasised in the organisations' strategies. Similar dynamics between the organisation—regarded as an open system—and its cultural contexts are likely to play out also in relation to other RRI keys and dimensions. Institutional characteristics that cut across nations are also an important influence: orientations towards certain visions of scientific excellence and researcher autonomy were common barriers to RRI implementation as identified in the workshops.

While the national mapping facilitates the understanding of the included research conducting and research funding organisations as open systems, micro studies have now started that will reveal more details about these organisations also as structural and natural systems. These studies will review the included organisations' work on the RRI keys, the AIRR dimensions and potentially other responsibility-related dimensions the organisations prioritise. In close contact with the organisations, Outlooks with concrete action plans and indicators for prioritized RRI aspects will be developed. These will form the basis for comparative research later in the project.

References

Boyle JP, DuBose ER, Ellingson SJ, Guinn DE, McCurdy DB (2001) Organizational ethics in health care: principles, cases, and practical solutions. JosseyBass, San Francisco

Brom FWA, Chaturvedi S, Ladikas M, Zhang W (2015) Institutionalizing ethical debates in science, technology and innovation policy: a comparison of Europe, India and China. In: Ladikas M, Chaturvedi S, Zhao Y, Stemerding D (eds) (Hrsg): science and technology governance and ethics: a global perspective from Europe, India and China. Heidelberg u.a.: Springer Open, pp 9–23

Cashore B (2002) Legitimacy and the privatization of environmental governance: how non-state market-driven (nsmd) governance systems gain rule-making authority. Gov An Int J Policy Administration Institutions, 15(4), 503–529

Forsberg E-M (2012a) Standardisation in the field of nanotechnology: some issues of legitimacy. Sci Eng Ethics 18(4):719–739

Forsberg E-M (2012b) Applying instruments for regional innovation—generating projects or legitimacy? Int J Innovation Reg Develop 4(5):430–445

Forsberg E-M, Eidhamar A, Kristiansen S-T (2012) Organising ethics: the case of the Norwegian army. Nordic J Appl Ethics 1:72–87

Hardy C, Maguire S (2008) Institutional entrepreneurship. In: Greenwood R, Oliver C, Sahlin K, Suddaby R (eds) The SAGE handbook of organisational institutionalism. SAGE, pp 198–217

Henisz WJ, Zelmer BA (2003) Legitimacy, interest group pressures and institutional change: the case of foreign investors and host country governments. In William Davidson institute working paper number 589. http://deepblue.lib.umich.edu/bitstream/2027.42/39975/3/wp589.pdf. Accessed 1 Mar 2011

Jacob K (2013) Options for strengthening responsible research and innovation: report of the expert group on the state of the art in Europe on responsible research and innovation. European Commission

Macnaghten P, Owen R (2011) Good Governance Geoengineering. Nature 479:293

Macnaghten P, Owen R, Stilgoe J, Wynne B, Azevedo A, de Campos A, Chilvers J, Dagnino R, di Giulio G, Frow E, Garvey B, Groves C, Hartley S, Knobel M, Kobayashi E, Lehtonnen M, Lezaun J, Mello L, Monteiro M, Pamplona J, Rigolin C, Rondani B, Staykova M, Taddei R, Till C, Tyfield D, Wilford S and Velho L (2014) Responsible innovation across borders: tensions, paradoxes and possibilities, Journal of Responsible Innovation, 1: 191–199

Moodysson J, Zukauskaite E (2012) Institutional conditions and innovation systems: on the impact of regional policy on firms in different sectors. In Papers in innovation studies 2011/13, Lund University, CIRCLE—Center for Innovation, Research and Competences in the Learning Economy

Owen R, Goldberg N (2010) Responsible innovation: a pilot study with the UK engineering and physical sciences research council. Risk Anal 30:1699–1707

Scott RC (1987) Organizations: rational, natural, and open systems, 2nd edn. Prentice Hall, Upper Saddle River, NJ

Von Schomberg R (2013) A vision of Responsible Research and Innovation. In Owen R, Heintz M, Bessant J (eds) Responsible innovation. London: John Wiley

Wittgenstein L (1953) Philosophical investigations. Blackwell Publishing

Chapter 2
Case Studies for Responsible Innovation: Lessons from Fukushima

Yuko Fujigaki

Abstract This paper deals with the NPP (nuclear power plants) accidents in Fukushima in 2011 in the context of responsible innovation. Based on our book, "Lessons from Fukushima: Japanese Case Studies on Science, Technology and Society", published by Springer in 2015, this paper examines the countermeasures to triple disasters (earthquake, tsunami, and NNP accidents) by the Japanese government. Focusing on the National Diet Report, I will show the recommendation by the report in 2012 and will introduce the reaction reports in 2017 by Cabinet Office to this report. After these analyses, I look into these recommendation and reaction in the context of RRI. Through these analyses, this paper shows one example of governance, public engagement and inclusion in the responsible R&D and Innovation process, dealing with Fukushima NPP accidents.

2.1 Introduction

An earthquake on March 11, 2011, triggered a large tsunami along the east coast of Japan, which damaged the cooling system of the Fukushima-Daiichi nuclear power plant and led to a hydro-explosion of the plant's core. The triple disasters hit Japanese Society and NPP accident has drawn the world's attention to the relationship between science technology and society in Japan.

To analyze the causes of disasters, three reports had been published in 2012. The first one is by National Diet, the second one is by Cabinet Office, and the third one is by an independent organization. Science Council of Japan also published several reports on de-contamination, temporary storage, reexamining an evacuated area, recompense, reconstruction support. In addition, to answer questions which I received from my colleagues in STS (Science and Technology Studies), I edited a book: "Lessons from Fukushima: Japanese Case Studies on Science, Technology and Society" (Fujigaki 2015). The book includes: historical analysis on how the

Y. Fujigaki (✉)
The University of Tokyo, Tokyo, Japan
e-mail: fujigaki@idea.c.u-tokyo.ac.jp

F. Ferri et al., *Governance and Sustainability of Responsible Research and Innovation Processes*, SpringerBriefs in Research and Innovation Governance,
https://doi.org/10.1007/978-3-319-73105-6_2

nuclear power plants are embedded in political, economic, and social contexts of Japan, analysis on communication failure between experts and public as well as reports on deliberative poll and public engagement for decision-making on how to deal with future energy.

Thus, many reports and statements have been produced for these 6 years on triple disasters. Do these reports have some effects to change the Japanese society? This paper verifies the countermeasures to these disasters by Japanese government focusing on the National Diet Report. First, I will show the recommendation by the report in 2.2, and then will introduce the reaction reports in 2017 by Cabinet Office to National Diet Reports in 2.3. After these analyses, I look into these recommendation and reaction in the context of RRI in 2.4.

2.2 Recommendation by National Diet Report

The National Diet Report (2012) proposed 7 recommendations based on the analysis on the NPP accidents as follows.

- **Recommendation 1: Monitoring of the nuclear regulatory body by the National Diet**
  A permanent committee to deal with issues regarding nuclear power must be established in the National Diet in order to supervise the regulators to secure the safety of the public.
- **Recommendation 2: Reform the crisis management system**
  A fundamental reexamination of the crisis management system must be made. The boundaries dividing the responsibilities of the national and local governments and the operators must be made clear.
- **Recommendation 3: Government responsibility for public health and welfare**
  Regarding the responsibility to protect public health, long-term monitoring must be implemented as soon as possible.
- **Recommendation 4: Monitoring the operators**
  TEPCO must undergo fundamental corporate changes, including strengthening its governance. In order to prevent the Federation of Electric Power Companies (FEPC) from being used as a route for negotiating with regulatory agencies, new relationships among the electric power companies must also be established—built on safety issues, mutual supervision and transparency.
- **Recommendation 5: Criteria for the new regulatory body**
  The new regulatory organization must adhere to the following conditions. It must be: Independent, Transparent, Professional, Consolidated, and Proactive.
- **Recommendation 6: Reforming laws related to nuclear energy**
  Laws concerning nuclear issues must be thoroughly reformed.

- **Recommendation 7: Develop a system of independent investigation commissions**
  A system for appointing independent investigation committees, including experts largely from the private sector, must be developed to deal with unresolved issues, including, but not limited to, the decommissioning process of reactors, dealing with spent fuel issues, limiting accident effects and decontamination.

2.3 Report in 2017 by Cabinet Office

How have the government reacted to this recommendation by National Diet? It has been past 6 years from the accidents and the Cabinet office summarized the 2017 report and submitted to the No. 193 regular session of the Diet on June 16th, 2017. The reactions to the recommendations are as follows.

- **To Recommendation 2: Reform the crisis management system**
  The National Diet revised Act on Special Measures concerning Nuclear Emergency Preparedness on September 2012, and based on it the government expanded task force on Nuclear Emergency Preparedness. In addition, the National Diet revised Atomic Energy Fundamental Act on September 2012, and based on it the government established Nuclear Emergency Preparedness committee for coordinating measures to prepare for an emergency in the ordinary times.
- **To Recommendation 3: Government responsibility for public health, etc.**
  The government provides subsidy to the "Public health funds for Fukushima residents." Using this funds, the Fukushima prefecture conducted Health Survey for residents. It also makes materials applying UNSCEAR (United Nations Scientific Committee on the Effects of Atomic Radiation) publication titled 'Levels and effects of radiation exposure due to the nuclear accident after the 2011 great east-Japan earthquake and tsunami' which is published April 2014. In addition, the government conducted countermeasures on de-contamination, temporary storage, reexamining an evacuated area, recompense, reconstruction support.
- **To Recommendation 4: Monitoring the operators**
  The Nuclear Regulation Committee set up the plan for transparency to monitor the operators of NPP (e.g. TEPCO). JANSI (Japan Nuclear Safety Institutes) evaluates the operators like TEPCO, makes proposals, and supports operators to pursue the highest safety level in the world.
- **To Recommendation 5: Criteria for the new regulatory body**
  The government established Nuclear Regulation Committee in September 2012 set down by article 3 of the National Government Organization Act. The establishment law of this regulation committee specifies the duty, how to exercise the committee's authority, organization, decision, etc. There are 5 principles:

Independent decision making, effective exercise, open and transparent organization, aspiring and responsibility, and rapid-response in emergency.

- **To Recommendation 6: Reforming laws related to nuclear energy**
  Atomic Energy Fundamental Act set the main objective of Nuclear energy as to contribute citizen's health and safety. Regulations on the atomic reactor are relegated from Electricity Business Act to the Nuclear Reactor Regulation Law.

As for Recommendation 1 and 7, there are no comments in the report by Cabinet office. However, the National Diet established the Select Investigation Commission on Nuclear Power concerning **Recommendation 7**. In this commission, Advisory Board was established on May 2017, concerning **Recommendation 1.**

2.4 Evaluation from the Point of View of RRI

The reaction report by the Cabinet Office focused on "independency, effective exercise, open and transparent organization, aspiring and responsibility, and rapid-response in emergency" as we see in the countermeasures for Recommendation 5. The Cabinet Office also focused on subsidies for health survey and countermeasures on de-contamination, temporary storage, etc., in Recommendation 3. How can we verify these reactions from the point of view of RRI?

The essence of the RRI is considered as (1) Open up questions, (2) Mutual discussions, and 3) New-institutionalization. From these points of view, Cabinet office report fails to open up questions, to enhance mutual discussions, and to exercise new-institutionalization, even though it considers independency and transparency. For example, as for **Recommendation 3**, to build a mutual trust for conducting a health survey for residents, doctors in Fukushima Medical University had held IAEA-FMU international conference for several times; however, there still exists segregation among residents and distrust to doctors and to governments (Fujigaki 2015). Furthermore, for resolving this distrust, "opening up questions" and "mutual discussions" are important. There are several trials to include citizens for discussion in health, but not in success. For setting a successful citizen conference on health survey, we have to deal with segregation among residents.

At the same time, the government declares that they seek "the highest safety level in the world" (e.g. to **Recommendation 4**), but criteria for this highest safety level were decided in closed space by experts: there is no space for reflecting residents' voice, and for opening up discussions and mutual discussions. For example, the gap exists between experts and residents on "what is safety"; however, the government had little effort to fill this gap.

In addition, the government should analyze the structure which invites distrust and has to reflect this analysis to next new-institutionalization. However, for establishment of the new regulatory body (**Recommendation 5**), there are little trials to set up space to open questions and mutual discussions for considering new-institutionalization of regulation.

How can we deal with these situations? What are measures to change Japanese society from "paternalism" to a "democratic society with public engagement"? Paternalism in Japanese society means that people rely extensively on professionals. Extensive public trust in professionals co-exists with a lack of democratic control over the nuclear power. On this point, the American historian T. M. Porter, who studied severe public scrutiny in flood control in the U.S., indicated that the "Japanese nuclear engineers were insulated to a striking degree from public scrutiny of the sort faced by American ones" (Porter 2013). Lack of democratic control and of public scrutiny in Japan was exposed to the light of day by the accident.

For constructing new relationships between citizens and professionals without extensive reliance on professionals, with democratic control, and with public scrutiny, the essence of RRI, open up questions, mutual discussions, and new-institutionalization is inevitable. Nuclear engineers have a sense of crises and do not share a rosy, optimistic picture of nuclear power by Ministry of Economic, Trade and Industry. In addition, The Fukushima nuclear power plant accidents and the Japanese DP attempt have invoked many international reactions. Japanese policy and citizen movements seem to be attracting worldwide attention. Therefore, for the responsible R&D, the Japanese Nuclear engineering should consider how to include the essence of RRI to establish NPP in the world.

2.5 Conclusion

From the analysis of the Recommendations by the National Diet and of reaction report by Cabinet Office, it was found that Japanese Cabinet office report fails to open up questions, to enhance mutual discussions, and to exercise new-institutionalization. The remaining problems are: (1) how to establish the space for opening questions and mutual discussions to rebuild trust in professionals without extensive reliance on professionals and with public engagement in Japanese society; and (2) based on (1), how to institutionalize new regulatory bodies to construct new relationships between science, technology, and society with democratic control and with public scrutiny in Japan.

References

Cabinet office investigation committee on the accident at the fukushima nuclear power stations (2012). Final report. Retrieved 5 August 2014, http://www.cas.go.jp/jp/seisaku/icanps/eng/final-report.html

Fujigaki Y, Tsukahara T (2011) STS implication of Japan's 3/11 crisis. East Asian Sci Technol Soc 5(3):381–394

Fujigaki Y (2015) (eds) Lessons from fukushima: Japanese case studies on science, technology and society. Springer

Independent investigation commission on the fukushima daiichi nuclear accident (2012). Tokyo: Discover 21

National Diet (2012) National diet official report of Fukushima nuclear accident independent investigation commission). Tokyo: Tokuma shyoten

Porter TM (1995) Trust in numbers: the pursuit of objectivity in science and public life. Princeton University Press, Princeton

Porter TM (2013) Preface for the Japanese edition of trust in numbers. Tokyo: Misuzu syobo

Chapter 3
A Report from the Field: Doing RRI from Scratch in an Assisted Living Technology Research and Development Project

Ellen-Marie Forsberg and Erik Thorstensen

Abstract The transdisciplinary Assisted Living project conducts research within ICT, health science, social science and ethics. The overall aim of the project is to advance responsible research and innovation (RRI) in the field of welfare technology. By adapting an RRI framework, the project aims to: (a) map how stakeholders and experts perceive the state-of-the-art of responsible welfare technologies, focusing on assisted living technologies (ALT), in Norway and internationally; (b) develop ALT solutions for users with mild cognitive impairment and dementia (MCI/D), through an RRI approach; (c) judge by an integrated HTA approach whether technologies introduced through an RRI process score better than currently implemented technologies; and d) create a wider dialogue on responsible welfare technologies for the future, reflecting on alternatives and options. In the project RRI is operationalized as involving four dimensions: (i) A specific focus on addressing significant societal needs and challenges, (ii) A research and development process that actively engages and responds to a range of stakeholders, (iii) A concerted effort to anticipate potential problems, identify alternatives, and reflect on underlying values, and (iv) A willingness from relevant actors to act and adapt according to 1–3. These dimensions are built into the project's design in different ways. The project, funded by the Research Council of Norway, started December 2015 and we have by now had substantial experience with working with these dimensions in practice. This paper will describe the experiences with including needs assessment, engagement, anticipation, reflection and responsiveness in the project, after 1.5 years operation. The paper will highlight several challenges that

---

The original version of this chapter was revised: For detailed information please see Erratum. The erratum to this chapter is available at https://doi.org/10.1007/978-3-319-73105-6_13

E.-M. Forsberg (✉) · E. Thorstensen
Work Research Institute, Oslo and Akershus University College, Oslo, Norway
e-mail: ellenmarie.forsberg@hioa.no

E.-M. Forsberg · E. Thorstensen
Akershus University College (HiOA), Pilestredet 35, 0166 Oslo, Norway

F. Ferri et al., *Governance and Sustainability of Responsible Research and Innovation Processes*, SpringerBriefs in Research and Innovation Governance,
https://doi.org/10.1007/978-3-319-73105-6_3

have appeared in the project when doing RRI in practice, related to transdisciplinarity, communication, project planning and control, and quality. We believe that the challenges experienced in our project are typical of RRI projects, so it is important to create open discussions about the pros and cons of RRI projects in the community of RRI practitioners.

3.1 Introduction

Assisted living technology (ALT) is a generic term for a heterogeneous group of technologies, often used in care for persons with mild cognitive impairment or dementia (MCI/D), involving for example videophones, robotics, GPS technology and monitoring systems to enhance security and safety and enable people to live an independent everyday life at home and in the community. The political interest in ALT has been more pronounced than the actual use of these technologies, because of factors related to organisational culture among care providers, technological alienation among elderly, and a lack of anchoring of the technologies with the relatives of the persons with MCI/D (Calvaresi et al. 2017). In contrast, within a Responsible Research and Innovation (RRI) line of thought innovations should not be pushed *on* society, but rather be developed *with* society, to meet the needs *of* society. RRI has been interpreted as a comprehensive approach of proceeding in research and innovation in ways that allow all stakeholders at an early stage (A) to obtain relevant knowledge on the consequences of the outcomes of their actions and on the range of options open to them and (B) to effectively evaluate both outcomes and options in terms of societal needs and moral values and (C) to use these considerations (under A and B) as functional requirements for design and development of new research, products and services (European Commission 2013a: 3).

However, even if there seems to be a need for RRI in ALT development, practising RRI raises some specific challenges in this field. Many of the main stakeholders (persons with MCI/D) will have problems discussing technology options with researchers and developers. An RRI approach will, therefore, have to adapt to this situation by facilitating communication at a level on which the user is able and comfortable to engage, and also involve other relatives or supports the user may have. Another specificity of RRI in assisted living technologies is an apparent technological alienation among the main users.

The still rather philosophical concept of RRI has been operationalised in the ICT field through projects such as *FRRIICT*, *ETICA* and *Framework for RRI in ICT,* and specifically for assisted living technologies in the *Responsible Industry* project. However, there is as of yet no commonly agreed upon procedure or governance framework for RRI, neither for research and innovation in general, for ICTs nor for assisted living technologies. There is thus a need for case studies and experiments with incorporating RRI approaches in technology development in this field, and the Assisted Living project amounts to an important contribution to such experimentation.

3.2 A Case Study in RRI

The Assisted Living project engages in development of technological solutions to needs defined by persons with MCI/D themselves in an RRI process specifically adapted to the capacities of such user groups.[1] By adapting an RRI framework, the project aims to: (a) map how stakeholders and experts perceive the state-of-the-art of responsible welfare technologies, focusing on assisted living technologies, in Norway and internationally; (b) develop ALT solutions for users with mild cognitive impairment and dementia, through an RRI approach; (c) judge by an integrated HTA approach whether technologies introduced through an RRI process score better than currently implemented technologies; and (d) create a wider dialogue on responsible welfare technologies for the future, reflecting on alternatives and options. The project is designed to be both an important contribution in the process of ALT innovation in Norway and to enrich and inform RRI in both concept and practice. The project is transdisciplinary and integrated, with project partners from nursing science, occupational therapy, automation and electronic engineering, sociology, philosophy, ethics and technology assessment (TA), and includes a smart home/welfare technology company and close collaboration with Oslo Municipality.[2] The overall aim of the project is to advance responsible research and innovation (RRI) in the field of welfare technology.

The technology intervention in the project is carried out in a housing complex in Oslo, where elderly that have certain needs, but are still able to live independently at home, can rent apartments in a building that also contains a restaurant, physiotherapy, exercise groups, etc. On the generic level, the technology solutions explored in the project will provide an automated environment to support the patient's everyday activities and provide a framework of safety. We incorporate existing sensor, telecommunication and automation technology and develop self-learning solutions that interpret, anticipate and intervene as required. The specific nature of the solutions explored in the project is determined as a result of user engagement and is therefore still under development. However, the long-term goal of the project is to develop self-learning systems (i.e. machine learning) that can provide useful cognitive support in accordance with individual values, choices, and needs.

This paper will give a brief description of the RRI method in the Assisted Living project's technology development project. The project started up in December 2015 and runs over four years, so we here only present the RRI design of the project and some preliminary results.

---

[1]See https://assistedlivingweb.wordpress.com/. The project is funded by the Research Council of Norway, under the SAMANSVAR programme (grant no 247620/O70).

[2]See https://assistedlivingweb.wordpress.com/english/partners/.

3.3 Operationalising RRI

In the project we refer to Wickson and Forsberg's (2013) spelling out of what they argue to be common dimensions in most RRI approaches:

For research and innovation to be responsible it needs to include:

1. A specific focus on addressing significant societal needs and challenges,
2. A research and development process that actively engages and responds to a range of stakeholders,
3. A concerted effort to anticipate potential problems, identify alternatives, and reflect on underlying values, and
4. A willingness from relevant actors to act and adapt according to 1–3.

These four dimensions are then operationalised in different ways. In the remainder of the paper we will go through these four dimensions and present some preliminary reflections on their operationalisation in practice.

3.4 Addressing Significant Societal Needs and Challenges

The project inherently addresses the grand challenge of the aging population. In addition, we have designed a comprehensive process to better understand the needs of elderly. We do this in several ways;

1. a survey among elderly who receive home-based services
2. open dialogue cafés with elderly at the study location in Oslo
3. inclusion of a selection of elderly from the study location in a technology intervention study, including a thorough assessment of their needs
4. focus groups with employees in home based services
5. focus groups with next-of-kin to the individuals included in the technology intervention

At this point, we already have quite a lot of results from activities a) to d), but we still include more elderly in the survey, will organise more dialogue cafés and will include more individuals in the technology trials.

An important challenge so far has been to uncover real needs. We have so far organised four dialogue cafés where we first explored general challenges in elderly's daily lives, then explored generic technological solutions to some of these challenges, proceeding to explore pros and cons of some concrete solutions and finally to invite the participants to try out two solutions. However, even if eight participants were willing to try out these solutions, it turns out that several of them do not actually have much need for them.

There are several ways to account for this situation. One important reason is that the elderly at this location want to be positive and collaborate with us even if they don't personally need the solutions. Another might be that in the design of the dialogue cafés we asked the elderly to discuss generic user stories in order to avoid creating socially awkward situations where too much personal information was shared. This has resulted in feedback on generic situations, and not necessarily the individuals' particularities. We have also asked the elderly to be co-researchers with us, rather than insisting on them having certain needs we will solve in the project. This is an empowering research strategy, but does not guarantee that the project meets actual needs of specific individuals (even if there is evidence that elderly in general may have such needs). So in the project, we have uncovered challenges that probably are reasonably representative (see Zouganeli et al. 2017 for a list of these), but the first technology solutions to be tested do not necessarily represent solutions to needs experienced by this particular test group.

3.5 A Research and Development Process that Actively Engages and Responds to a Range of Stakeholders

In the project, we engage with elderly at the study location through dialogue cafés and technology trials, as described above. We also meet them through the survey in the home-based services. Through focus groups, we have also engaged with staff in the home-based services and we will engage with next-of-kin. These interactions help us outline the direction for the technology development in the process.

In addition to this, we engage with a broader range of experts and stakeholders to discuss the design of the research project as such, methodological choices, project activities and preliminary results. This group, called the ProjectSTEP group, functions as a combination of a steering group and a sounding board. An important function of the group is to discuss the situation analysis in the project and critically reflect on the framing of the issues and the plans and methods for developing and assessing the ALT solutions in the project. The group follows a version of the procedures of the so-called TranSTEP group as described in https://transtepapproach.wordpress.com/doing-transtep/ (see Forsberg et al. 2015).

Finally, we want the learning generated from the reflections in the project team, the ProjectSTEP group and dialogue cafés to be disseminated to and discussed with a larger group of national stakeholders. This will take place in two conferences; one focused on foresight and one on presenting and discussing the project's main learning points.

3.6 A Concerted Effort to Anticipate Potential Problems, Identify Alternatives, and Reflect on Underlying Values

As described above the mandate of the ProjectSTEP group is to help the project anticipate problems, identify alternatives and reflect on underlying values. In addition, such reflection has a dedicated slot in each consortium meetings' agenda. In this slot we specifically reflect on the learning processes in the project. In the kick-off meeting, we included the following reflection:

1. Write down 3 words that represent good transdisciplinary research cooperation for you
2. Any comments? Anyone who wants to explain their choice of words? Or comment upon others' choices?
3. How can we—and each one of us—ensure that these qualities are followed up in the project?

In the second consortium meeting we reflected on how much the different disciplines of the Assisted Living project need to understand of other disciplines and parts. Each team member was to fill in a three column set-up indicating what they believed they needed to know about other partners' research, what they would like to know and what they believed was not necessary for them to know.

In the third consortium meeting the reflective exercise was for each to fill in the following statement: When we present our solutions (as you envisage them) to the elderly, I expect [...] because [...]. The intention here was to explicate our assumptions and expectations about the elderly's relation to technology. As all project team members filled this out, it allowed us to better understand similarities and differences in assumptions and expectations within the consortium and reflect on these.

The fourth consortium meeting was a shorter meeting addressing a situation where it had become clear (referring back to the second reflection exercise) that some issues were elevated from being 'nice to know' to 'need to know', as they had implications for privacy questions, general research ethics and costs. The character of the technology research was here in focus and we discussed what characterizes RRI projects compared to other user-oriented research and development projects. A relatively unique feature of the Assisted Living project is that it is a technology development project lead by an RRI partner. In most integrated technology projects, the project leader is the technology partner and the RRI partner contributes in different ways to make the technology development process more reflective or responsive to societal concerns. In the Assisted Living project the balance between the partners—the technologists, the health researchers and the RRI partners—is tipped more in the direction of RRI, not least because there are four partners with RRI competence in the consortium.

These discussions are crucial, but challenging, as all partners need to reflect on their own roles, their own expectations to the project, assumptions about scientific quality, and wishes for project outcomes.

3.7 Responsiveness—A Willingness from Relevant Actors to Act and Adapt According to 1–3

The final dimension in the project is responsiveness. The project is designed to be responsive, and the funder,—the 'co-responsibility' program in the Research Council of Norway—has allowed for (and indeed encourages) flexibility to proposals from stakeholders during the project. An obvious expression of responsiveness is the way we have made decisions on technology solutions as a result of listening to the users.

Another expression of responsiveness is the project's policy with regard to input from the ProjectSTEP group. During the meetings in the ProjectSTEP group, the project team is not supposed to respond immediately, but rather listen (and explain, when this is necessary). After the ProjectSTEP meetings, the project team discusses what we've learned and responds systematically to the input. Some input we take simply for our information, other we adapt to, and some input we deem out of scope for the project. These responses are then published on the project's webpages (https://assistedlivingweb.wordpress.com/the-projectstep-group/), in order to transparently show how the project is responsive to input. An example of a change made from input from the ProjectSTEP group was to expand the project's focus from the MCI/D diagnosis to frail elderly in general, as the technological research in the project is not only relevant for those with a diagnosis.

In addition to these planned responsiveness measures, we have also learned the necessity of being responsive to unexpected practical issues that affect the research, especially related to the engagement of the elderly at the intervention site. In order to be successful, the project must be sensitive to social dynamics at the site, to practical particularities of the individual elderly's home, daily routines, varying health conditions and wishes, and to technical challenges related to sensor equipment, data transfer, etc. This makes it clear that the flexibility of such an intervention project is crucial. This flexibility also implies that ethical considerations cannot simply be carried out ex-ante, at the planning stage, but must be a continuous reflection as the project evolves.

3.8 Concluding Remarks

We are still in the very beginning of the technology trials in the process. These trials will give us more quantitative and qualitative data on de facto technology use among the elderly. Working with technology implementation in the field will also likely bring up further issues concerning the transdisciplinary interaction in the project. These results will be discussed as the project advances.

References

Calvaresi D et al (2017) Exploring the ambient assisted living domain: a systematic review. J Ambient Intell Humaniz Comput 8(239–257):239

European Commission (2013) Options for strengthening responsible research and innovation report of the expert group on the State of Art in Europe on responsible research and innovation. http://ec.europa.eu/research/science-society/document_library/pdf_06/options-for-strengthening_en.pdf

Forsberg E-M, Ribeiro B, Heyen NB, Nielsen RØ, Thorstensen E, de Bakker E, Klüver L, Reiss T, Beekman V, Millar K (2016) Integrated assessment of emerging science and technologies as creating learning processes among assessment communities. Life Sci Soc Policy 12:1–20. https://lsspjournal.springeropen.com/articles/10.1186/s40504-016-0042-6

Wickson F, Forsberg E-M (2014) Standardising Responsibility? The significance of interstitial spaces. Sci Eng Ethics 21:1159–1180

Zouganeli E, Casagrande FD, Holthe T, Lund A, Halvorsrud L, Karterud D, FlakkeJohannessen A, Lovett H, Kjeang Mørk S, Strøm-Gundersen J, Thorstensen E, Norvoll R, ter Meulen R, Kennedy M-R, Owen RJ, Ladikas M, Forsberg E-M (2017) Responsible development of self-learning assisted living technology for older adults with mild cognitive impairment or Dementia. In: Röcker C, O'Donoghue J, Ziefle M, Maciaszek L, Molloy W (eds) Proceedings of the 3rd international conference on information and communication technologies for ageing well and e-Health,, ScitePress. pp 204–209 https://doi.org/10.5220/0006367702040209

Chapter 4
Responsible Research and Innovation in Open Health and Open Science. Open Science: Open and Toll-Free Data Age Open Health: From P2P (Patient to Physician) to P2E (Person to Eco-System)

Fulvio Ananasso, Sergio Farruggia, Renzo Provedel and Monica Sebillo

Abstract Some specific case studies are addressed—Open Health and Open Science—where Responsible Research and Innovation can play a crucial role and yield potential benefits, aiming at openness, transparency and sharing of relevant science and innovation efforts and outcomes. Open Science concerns access to outcomes of Science, circulation and use of results, development of scientific findings and breakthroughs. It implies *open access*, *open data*, *open peer review* and *open licenses*, which influence copyright, patents and transfer of knowledge. We explore a "Country case" (Italy). Open Health includes epidemiology, health data of geo-localized populations, open innovation, caring diseases. Here we deal with the topic "care". Some people go beyond the patient-doctor exclusive relationship through unexpected resources and experiences, sharing medical records with the "crowd" to get ideas, perspectives, cures, experiences. A new paradigm by Salvatore Iaconesi is analyzed.

F. Ananasso (✉) · S. Farruggia · R. Provedel · M. Sebillo
Stati Generali dell'Innovazione, Rome, Italy
e-mail: fulvio.ananasso@alice.it

F. Ananasso
CDTI, Rome, Italy

S. Farruggia · M. Sebillo
AMFM GIS Italia, Rome, Italy

R. Provedel
FareImpresa, Genoa, Italy

M. Sebillo
University of Salerno, Fisciano, Italy

F. Ferri et al., *Governance and Sustainability of Responsible Research and Innovation Processes*, SpringerBriefs in Research and Innovation Governance,
https://doi.org/10.1007/978-3-319-73105-6_4

4.1 Introduction

The paper elaborates on some specific case studies under assessment at "Stati Generali dell'Innovazione" (SGI), a not-for-profit association established in 2011 with the goal of facilitating a shared perspective for innovation policies in Italy. The SGI scope of work, promoting collaboration among scientists, societal actors, Institutions, the political sector, citizens,… is totally in line with the Responsible Research and Innovation concept, part of the Europe 2020 strategy, aiming at achieving a wider dimension of science and innovation, improving the role of society in environmental/welfare state preservation, its socio-economic development and striving for a less unequal world.

In such a contest, some Open Science and Open Health topics are sketched in this paper, concerning specific initiatives related to (and requiring) openness, transparency and sharing of relevant science and innovation efforts and outcomes.

Open Science refers to such specific rules of access to the outcomes of Science so as to facilitate diffusion, utilization and further development of scientific findings/breakthroughs.

Under the Open Health definition we assume three different (but related) domains, i.e. (i) epidemiology (access to data relating to health of populations "geo-localized" across territories), (ii) care (new paradigm whereby the person suffering from a syndrome can bring the disease back to a social dimension, subtracting both disease and patient from isolation, code of conduct and language of specialists/healthcare providers) and (iii) open innovation—new paradigm to find "technical" solutions (e.g. care protocols, vital parameter monitoring, etc.) and "technology" solutions (e.g. vaccines, drugs. etc.) to a health problem by accessing worldwide-scattered knowledge.

In the paper we address in some detail the topic "care".

Open Science

4.2 Definitions

- **open access**: open and toll-free access and circulation of scientific documents, overcoming copyright (Pievatolo and Kant 2011);
- **open data**: free sharing of experiments data—"FAIR" principles (Wilkinson MD 2016);
- **open peer review**: independent review of scientific publications;
- **open licenses**: breaking general rules applied to licenses—e.g. duration.

This "open mode" interacts deeply with the actual context of patents. India is paradigmatic: it produces most of the drugs to cure Hepatitis C, overcoming patents effective in the Western World (The economic times 2016).

4.3 A reference model for scientific publications

Today the standard protocols to publish scientific papers require Authors to select a high reputation Publisher—taking also care of "peer reviewers" unknown to the Authors-, the biggest ones forming de facto a 3-party oligopoly—Elsevier (Holland), Wiley (USA) and Springer (Germany). Authors assign (without remuneration) the copyright to the Publisher, i.e. one can read the article only paying the Publisher, buying publications/access to databases. On the other hand, publishing is necessary for Authors to be evaluated, from both career and funding of Research points of views. Main impacts of the above: worldwide circulation of information slowed down; cost (mainly borne by Universities) becoming unbearable; open dissemination/cross transfer of knowledge made harder; and so on.

4.4 Scenario

A new paradigm of "Open Science" appeared on the international scenario several years ago, but only the "internet revolution" enabled the change process towards new best practices:

- free exchanges between disciplines: borders are vanishing;
- cross transfer: new frontiers of understanding and modeling;
- open behavior of researchers and scientist: from "ego" to "eco" systems.

This change process produced some effects towards "openness":

- open access to scientific publications: cost-savings and quicker reading;
- open access to experimental data: quicker validation;
- open peer review: improved dialogue between Reviewers and Authors.

4.4.1 Open Access Sources on the Internet

Kazakh researcher Alexandra Elbakyan succeeded created in 2011 the "Sci-Hub" database and site (Sci-Hub), a virtual library of some 62 million papers, accessible quickly and easily by private citizens. In 2015 Elsevier filed and won a lawsuit against Sci-Hub alleging copyright infringement (Sci-Hub Alleging 2017).

Nevertheless, Sci-Hub, although blocked in some Countries, kept on developing its contents, supported by relevant sectors of the scientific and academic communities.

There are other open access disciplinary repositories, like "ArXiv" (Cern) for Physics, "RepEc" for Economic sciences, "Dash" (institutional archive of Harvard University), etc. They usually host the full version of papers.

4.4.2 Open Peer Review

We see today some limits to peer review of scientific publications, being:

- not transparent: selection and evaluation protected by anonymity;
- not a public conversation between experts;
- "conservative": slowing down and creating obstacles to innovators.

Sharing of peer review conversations over social networks may offer a "revolutionary" meaning, as it was during 17th century. E.g. T. Gowers won a Fields medal for Mathematics with Discrete Analyses (Gowers), an overlay journal which selects through (free of charge) peer review the papers available on ArXiv (Cern).

4.4.3 Open Data and Patents

Open data and patents represent a "paradox". Scientific publications should disseminate the outcomes of researches, whilst patents are focused on the industrial process, invention or model, to protect its commercial exploitation. Their relationship is not well addressed in the scientific arena and the potential conflict with "re-using" scientific publications is not even clarified. We see the paradox of patenting more and more, at the same time "opening" data. As noticed by Roberto Caso (Caso 2016), EU (Horizon 2020) policy is rather ambiguous, dictating default open access publishing, should intellectual property rights (IPR) not be applicable.

4.5 Some Cases of Excellence

Geneva's CERN is a rather well-known case of Open Science. CERN makes available its articles on the ArXiv archive (Cern), its experimental data on http://opendata.cern.ch/ and provides worldwide researchers with a tool to share information (Zenodo). As known, the "www" was invented at CERN and is free. Another pioneering project was Open Source Malaria, which shared not only the results of the research but also its processes, the Lab Notebooks.

4.6 SWOT Analysis in Italy

Strengths	Weaknesses
17 Universities agreed and published open access rules for scientific papers Excellent digital culture of new generations of researchers for sharing and free publication of experiments Culture of "sharing" is spreading all around	Most Universities do not practice open access Evaluation system of research/researchers still conditioned by large publishers Universities still slow to adopt to new open approaches
Opportunities	**Threats**
Open access is becoming a need due to University budget decrease	Lobbying by large publishers of scientific publications

Open Health

4.7 Scenario

Health is considered by the Society a confidential matter, subject to privacy rules. At the intersection of many disciplines, like Medicine, it has to respect information restrictions, due to "Ethical code of conduct" and Laws. Furthermore, the State, managing "National Public Health Service", collects great deals of data, delivering the main health services through Hospitals and Laboratories. Health data are not public, but can be used as "big data" to understand phenomena and take public interest decisions. A major role is then played by *data*, triggering new games. The Internet is changing the perspectives and is innovating the use of data.

4.7.1 Big Data

Public Administration collects huge amounts of data to understand and intervene on pandemics/viral diseases; to manage efficiently the nationwide health services and infrastructures; to manage drugs and disposal of dangerous waste. Archives report infectious diseases, handicapped birth rate, mortality taxonomy, air/water pollution, meteorological data, emissions by industry/location, etc.

4.7.2 Personal Data Sharing and New Paradigms

For hundreds of years, data regarding diseases of "patients" have been treated as a "highly confidential" matter, between the patient and the physician. The Internet changed this well-established praxis: people access the web to search for symptoms, diagnosis, syndromes, therapies,… resulting deeply influenced by those.

We present here new approaches to "disease management", inspired by the experiences and thoughts out of a specific life experience of Salvatore Iaconesi (Iaconesi and Persico 2016), unveiling new emerging views and behaviour of the "patient":

- access to private data: open format to share easily private health data;
- disease as a social event: many people are involved—a waterfall effect;
- participation to "my problem-disease": new social codes;
- disease as a *performance*: it might be played by the "patient", even in unexpected fields such as art, communication, education,…

People want to overtake the mere logic of "receiving a service" and perception of being a "number" to enter a wider relationship with "others" (Terzani 2004). *Sharing data* plays a crucial role, opening spaces beyond the patient-physician closed circuit.

4.7.3 Open Innovation for Health

Health is more and more an interdisciplinary issue, and even companies outside the "pharma" can today address the health sector, also thanks to 'cross sector' *Open Innovation* paradigm (Provedel 2007). Many sectors can generate solutions to improve health (Ninesights). Pharma, developing stem cells and genomics; energy, providing cleaner fuels; packaging, reducing plastics; electronics, developing wearable sensors integrated into the human body; chemicals, using nanotech for diagnostics; nutrition, improving quality of life and ageing. The general public is called to action as solution providers in this scenario—social networks being relevant enablers.

4.7.4 A New Paradigm on the Stage

Salvatore Iaconesi, who got sick of brain cancer in 2012, decided to change the game (Iaconesi 2013). He opposed the usual role of "patient", and decided to change the path of his disease tunnel, embracing a new paradigm, "the Cure", based on two pillars: (i) *sharing his disease data* with the world; (ii) *connecting to the world as a person* to get any support, idea, tool, solution, experience,… to fight against his disease.

***Stage zero*. *The disease*.** Salvatore Iaconesi, artist, hacker, teacher,… in 2012 gets a brain cancer diagnosis. He then wishes to start a dialogue with the disease and his community, contrasting obstacles to divest his role of "patient": everyone is talking *of* him and of his disease, not *with* him. Data are *not belonging to him*.

Stage 1. Starting up the "Cure". Salvatore translates data into easy-to-read formats, JPG and HTML, and posts online his clinical records, asking who can find a cure. Artists, researchers, scientists, students,… reply using their own language and knowledge. The new paradigm, "cure", is born. The confidentiality of data taboo is transgressed. Everyone is asked to "redefine" the meaning of "cure".

Stage 2. Action time: dialogue and connectivity. Thousands of people offer project mappings, artworks, dancing performances, 3D printings, poems, scientific papers, data, storytelling,… changing the meanings of "disease".

Stage 3. Surgery. 5 months after diagnosis, a success surgery removed cancer.

Stage 4. Long term deliveries. Solutions, devices, services by the "crowd" collaboration. The book "la Cura" has three sections: his storytelling; technical/ scientific issues; a toolkit that everyone can use to apply "the cure" to himself.

Stage 5. The future. Such methodology can be generalized to other diseases/ sectors (education, environment, wellness,…), going beyond the disease context.

4.8 Conclusions and Discussion—a New Emerging Paradigm for Health

Traditional models "patient-physician" can evolve to models "cure" as:

- Disease borders: from dyadic "patient-doctor" to networking. The definition of disease itself changes. "Socialization" do involve a variety of subjects: friends and relatives, researchers, artists, designers. A "network" appears around the patient. The cancer belongs to the participating network. Everybody can have a role and be able to bear it properly.
- Disease process: "stay in the flow". Being in the flow is much different from question and answers. Monitor and support, more than exchange mails.
- Disease, the "after". *continuum* is beyond the solution, long lasting and generating lifelong outcomes.

Data, social, connectivity are key ingredients to create a different scenario.

Acknowledgements The authors greatly appreciate the contribution by Salvatore Iaconesi (Open Health) and Maria Chiara Pievatolo (Open Science).

References

Caso R (2016) La scienza aperta contro la mercificazione del sapere. http://www.cittadellascienza.it/centrostudi/2016/04/la-scienza-aperta-contro-la-mercificazione-del-sapere/

Cern (2017). arXiv http://arxiv.org

Gowers T Discrete analysis (2017). http://discreteanalysisjournal.com/

Iaconesi S (2013) Cos'è successo quando ho reso open-source il mio cancro al cervello, TED. https://www.ted.com/talks/salvatore_iaconesi_what_happened_when_i_open_sourced_my_brain_cancer?language=it
Iaconesi S, Persico O (2016) La cura, Codici Edizione
Lab notebooks (2017). http://malaria.ourexperiment.org/
Ninesights (2017). https://ninesights.ninesigma.com/projects
Open source malaria (2017). http://opensourcemalaria.org/
Pievatolo MC, Kant I (2011) Sette scritti politici liberi, Firenze University Press. http://btfp.sp.unipi.it/dida/kant_7/index.xhtml#liberarekant
Provedel R (2007) L'innovazione aperta: come catturare le idee dal mondo attraverso i broker dell'Innovazione, Mondo digitale, n.1
Sci-Hub (2017). http://scihub.org/
Sci-Hub Alleging (2017). https://www.nature.com/news/us-court-grants-elsevier-millions-in-damages-from-sci-hub-1.22196
Terzani T (2004) Un altro giro di giostra, Longanesi
The economic times (2016) Patent not to affect hepatitis C drug price in India, Gilead Sciences Inc. http://economictimes.indiatimes.com/industry/healthcare/biotech/pharmaceuticals/patent-not-to-affect-hepatitis-c-drug-price-in-india-gilead-sciences-inc/articleshow/52228499.cms
Wilkinson MD et al (2016) The FAIR guiding principles for scientific data management and stewardship. https://www.nature.com/articles/sdata201618.pdf
Zenodo (2017). https://zenodo.org/

Chapter 5
Inclusive Business Models and RRI: Insights from the ProGReSS Project

Antonia Bierwirth and Raúl Tabarés Gutiérrez

Abstract The Inclusive Business models or commercially viable business models that incorporate poor people into their value chains, open up new growth opportunities for businesses, while simultaneously promote sustainable development. A growing number of large corporations are proactively engaging in such approaches with the aim of achieving core business benefits through improved societal desirability of their innovation products and services. Researches show that in order to create a Bottom of the Pyramid (BoP) market, companies should focus on a proposition of mutual value creation: The greater the ability of the enterprise to meet the needs of the poor, the greater the return to the partners involved (Antúnez-de-Mayolo 2012). The main problem in the context of societal desirability within business is that currently there is a lack of agreement about what constitutes social value in a globalized society and economy. To resolve this challenge, value should not created by firms acting autonomously, but by working together with local allies, and ideally by integrating them into the R&D and production chain. The mutual dependency that innovative businesses and Bottom of the Pyramids societies have on each other plays a central role in the Responsible Research and Innovation (RRI) framework and is explained in this paper through two concepts: Inclusiveness and Pro-Poor growth reducing inequality.

5.1 Reducing Inequality: A Greater Level of Prospective Global Responsibility

The Bottom of the Pyramid (BOP) is a socio-economic concept describing the world's poorest citizens who live with less than $2.50 a day and are excluded from the modernity of our globalised civilised societies (Financial Times Lexicon 2017). A study conducted with the IFC to understand the size and aggregate purchasing power of the BoP identified nearly 4 billion people at the BoP across Africa, Asia,

A. Bierwirth (✉) · R. T. Gutiérrez
Tecnalia Research and Innovation, Donostia/San Sebastián, Spain
e-mail: antonia.bierwirth@tecnalia.com

F. Ferri et al., *Governance and Sustainability of Responsible Research and Innovation Processes*, SpringerBriefs in Research and Innovation Governance,
https://doi.org/10.1007/978-3-319-73105-6_5

Eastern Europe, Latin America and the Caribbean (Hammond et al. 2007). Some estimates based on the broadest segment of the BOP put its demand as consumers at about $5 trillion in Purchasing Power Parity terms, making it a desirable objective for innovative businesses (Financial Times Lexicon 2017). Most companies trying to do business with the 4 billion people who make up the world's poor offer products at extremely low prices and margins, and hope to generate decent profits by selling enormous quantities of them. However, selling products to customers in rural villages and in slums near urban centres is difficult (Simanis 2012). Any business that starts off needing a 30% or higher penetration rate is built on a shaky foundation, because consumers at the bottom of the pyramid are not accustomed to using and experimenting with products and thus sales and marketing require expensive proposition (Simanis 2012). At the same time, in a very poor country, it is arithmetically impossible to reduce poverty without growth because there is no one to redistribute from (Commission on Growth and Development 2008) Developing countries can better grow with the support of the advanced economies as they need job creation, access to the open global trading system and opportunities to promote their exports until their economies have mature (Hammond et al. 2007). However, the contribution of the growth process to poverty reduction does not depend only on the rate of economic growth, but also on the ability of the poor to respond to the increasing demand for labour in the more productive categories of employment. Consequently, what matters is not just the rate of growth, but also the rate of poverty reduction for any given rate of growth, i.e. the growth elasticity of poverty. **Strategies for RRI aiming poverty reduction should promote both a faster rate of growth and a high growth elasticity of poverty** (UN General Assembly 2005). Currently, there are two main competing definitions for pro-poor growth—an absolute definition and a relative definition. Under the absolute definition growth is considered to be pro-poor as long as poor people benefit in absolute terms, as reflected in some agreed measures of poverty (Ravallion and Chen 2003). In the relative definition, growth is "pro-poor" only if the incomes of poor people grow faster than those of the population as a whole, i.e., inequality declines (White and Anderson 2000). The second concept is based on a relative concept of inequality and claims that inequality would fall with growth whenever growth is pro-poor. **Pro-poor measures within the framework of RRI should focus primary on reducing inequality as these measures provide a greater level of prospective global responsibility.**

5.2 RRI and the Need for Inclusive Growth

As argued above, growth alone does not guarantee that everyone shares equitably its benefits. Pro-poor growth is mainly concerned with impacts across the distribution of income, but not with the process of growth itself or with the marginalization of other groups not based on economic criteria, such a gender or race for example. Therefore, RRI involves another term of "inclusive growth" to complement the pro-poor growth concept. It takes a longer term perspective and refers to both the pace and the pattern of the economic growth (Hammond et al. 2007). Thus it relates to the realisation that growth processes may have different impacts not just across the distribution of income, but also among ethnic and gender groups and geographical regions. Rather than contemplating outcomes as the only important aspect, it takes into account whether and how people engage in the growth process itself (Ranieri and Ramos 2013). The main characteristics of Inclusive growth can be summarised as follows (Ianchovichina and Lundstrom 2009):

- **Equality of opportunity and non-discrimination:** The growth should be broad-based across sectors, and inclusive of the large part of the country's labour force. The inclusiveness refers to equality of opportunity in terms of access to markets, resources and unbiased regulatory environment for businesses and individuals (macro and micro determinants of growth) (Hammond et al. 2007). Thus it considers not only the firm, but also the individual as the subject of analysis. It also involves the notion of non-discrimination, meaning that no person or group is denied participation in the growth process (Klasen 2010). Thus, growth should benefit all stripes of society, including the poor, the near-poor, middle income groups, and even the rich (Klasen 2010).
- **Participation:** Inclusive Growth is about participation, not just outcomes. It allows people to contribute to and benefit from economic growth. It is both an outcome and a process: On the one hand, it ensures that everyone can participate in the growth process, both in terms of decision-making for organizing the growth progression as well as in participating in the growth itself; on the other hand, it makes sure that everyone shares equitably the benefits of growth (Ranieri and Ramos 2013. Most institutions also emphasise that this improved participation must also be explicitly linked to improved benefit-sharing, rather than assume that this will be automatic once opportunities are increased (CAFOD 2014).
- **Productivity:** Transfer and distribution schemes cannot be an answer in the long run and can be problematic also in the short run (Ianchovichina and Lundstrom 2009). The focus is not only on income and employment growth, but also on productive employment and productivity growth, which is a crucial source of growth in living standards. Productivity growth means more added value to production and more available income to be distributed.
- **Sustainability:** Inclusive growth is concerned with sustained growth, which means a constant and positive growth rate in capital per capita in a long term. It involves also the concept of sustainable prosperity that brings long-term benefits for environment, biodiversity and also contributes to the conservation of the natural resources.
- **Quality over quantity:** Growth is deemed to be inclusive, depending on the extent to which poor men and women have benefited through increased income, but the OECD 2013 also states that a priority should be given to the quality of growth over quantity of growth. Similarly, the UNDP emphasises the importance of putting human development first, especially sustainable human development achievements (UNDP 2010).

5.3 RRI Business Models

There are three different ways in which RRI actions in developing countries may become more inclusive to the Bottom of the Pyramid:

1. **Consumer-focused RRI**: Social challenges are solved through innovation being relevant to the poor; In this case the output is innovation that develops and delivers products and services improving low-income households' access to health, education, water and sanitation, food and nutrition, energy, housing, information, financial services, etc. Consumer-focused businesses are about far more than a cheap product; synergistic innovation is required across the whole business model and often along the entire value chain. The RRI pillar ethics is most prominent in the consumer-focused models. Ethics is the basic concept of respecting human dignity and aligning research to global social values.
2. **R&D focused RRI**: It is the process of innovation itself where the poor are actively engaged in some manner in the development of innovative solutions to their problems; these models best reflect the Pubic Engagement pillar of the RRI concept. R&D focused RRI actions promote the establishment of iterative and inclusive participatory multi-actor dialogues between researchers, policy makers, industry and civil society organisations, NGOs, and all stripes of society. They foster mutual understanding, co-creation of research and innovation outcomes. They also imply a shared risk of innovation and therefore enable wider acceptability of the results.
3. **Producer focused RRI**: The poor can participate in the production chain in different ways, for instance as employees, producers, suppliers, distributors, retailers or entrepreneurs. In all cases, they acquire the capacities to identify and absorb innovative solutions to their problems. This approach helps to create a skilled, adaptable workforce that is capable of competing in an increasingly knowledge-based and globalized marketplace. The Science Education Pilar of RRI plays here a crucial role here; it develops the poor´s competencies for problem-solving and innovation, as well as their analytical and critical thinking. On the long term it creates and nurtures an innovative new generation capable of understanding global problems as feeding and housing our population, healthy living, protecting our environment, generating sufficient energy, supplying enough clean water, urbanization and global climate change.

If the problems of the poor are defined and solved by outsiders, one can speak of innovation *for* the poor. If the poor are participating in the process this could be described as innovating *with and for* the poor (Simanis 2012). Consumer-focused RRI can solve social challenges but their impact is short term as they satisfy immediate needs of the poor. R&D focused RRI presents a significant advance as the poor co-participate in the innovation process. Co-creation implies joint creation of commodities and also a new understanding of buying as part of an experience rather than a mere transaction (Prahalad 2000). With this approach companies gain support from local formal and informal allies, face better the reputational challenges and most importantly, build legitimacy. Producer focused RRI have the highest inclusiveness level, they not only enclose the first two dimensions, but also suggest a participation in the decision-making and in the benefits sharing. They have a strong capacity building effect on strengthening the skills, competencies and abilities of people and communities in developing societies. As the poor form part of the value chain it is expected that their incomes grow faster than those of the population as a whole. Therefore, producer focused RRI actions have a greater chance than R&D focused RRI to produce pro-poor growth that reduces inequality.

5.4 Case Study from the FP7 ProGReSS Project: RRI with and for the Poor

The FP7 ProGReSS demonstrates that it is possible to co-innovate with poor using their valuable traditional knowledge in order to develop new marketable products. In the case study the collaboration between the San Community, an indigenous community located in Southern Africa, and HG&H an innovative pharmaceutical company, resulted in a successful marketable product Zembrin®, a natural remedy used against anxiety and depression. (Table 1)

The collaboration of traditional knowledge holders with pharmaceutical and other industries is an excellent example to illustrate the necessity and feasibility of RRI in low and middle-income countries, as it turns the poor into co-creators of the product but also gives them the possibility to participate in the production chain. Zembrin® is based on traditional plant knowledge of the Sceletium plant originating from the San community. The villagers' input on what plants to use, doses, and participation in the pharmaceutical studies were pivotal to the initial research and development on Zembrin®. In March 2013, Zembrin® was recognised as the

Table 1 Milestones in innovation based on San traditional knowledge (Cavallaro et al. 2014)

Year	Event
1986	Research by medical doctor/botanist begins based on literature on Sceletium plant
1995	Social science research involving community elders to exclude addictive properties
1996	Active component identified, patent obtained
2005	First standardised extract
2006	Halls Investment joins as a venture capitalist
2007	Contact made with San community for benefit sharing according to the Convention on Biological Diversity (1992)
2008	Benefit sharing contract concluded
Since 2007	Clinical studies
2010	Approval of the first product by South African Ministry (trade name - Zembrin)

"most sustainable ingredient" at the annual Natural Products Expo West held in Anaheim, USA, and was also noted as one of the leading ingredients pioneering the foundation for a new generation of natural products (HG&H Pharmaceuticals (Pty) Ltd 2013). Appropriately, the Zembrin® founders, HG&H Pharmaceuticals Ltd, signed Africa's first "prior informed consent benefit-sharing agreement" with the San indigenous community. In terms of this agreement they share 6% of all income from Zembrin® with the San Council which allocates 50% of their share with the communities of Paulshoek and Nourivier (HG&H Pharmaceuticals (Pty) Ltd 2013).

References

Antúnez-de-Mayolo C (2012) The role of innovation at the bottom of the pyramid in Latin America: eight case studies. Procedia—Soc Behav Sci. 40 http://www.sciencedirect.com/science/article/pii/S1877042812006398

CAFOD (2014) What is inclusive growth? CAFOD, London. http://www.cafod.org.uk/content/download/17223/133621/file/Inclusive%20Growth%20full%20paper.pdf

Cavallaro F, Obach M, Schroeder D, Chennells R, Snyders L, Steenkamp A, Bierwirth A, Kumar A (2014) Responsible research and innovation and end-users. Report for FP7 Project "ProGReSS", http://www.progressproject.eu

Commission on Growth and Development (2008) The growth report: strategies for sustained growth and inclusive development. World Bank, Washington, DC

Financial Times Lexicon (2017) http://lexicon.ft.com/Term?term=bottom-of-the-pyramid-(bop)

Hammond AL, Kramer WJ, Katz R, Tran J, Walker C (2007) The next four billion: market size and business strategy at the base of the pyramid. World Resources Institute and International Finance Corporation, Washington, DC

HG&H Pharmaceuticals (Pty) Ltd (2013) South African company set to revolutionise global stress-reduction market (20 March 2013). http://www.zembrin.com/media-centre/media-resources/press-releases

Ianchovichina E, Lundstrom S (2009) What is inclusive growth? the world bank (10 Feb 2009)

Klasen S (2010) Measuring and monitoring inclusive growth: multiple definitions, open questions, and some constructive proposals. In: ADB sustainable development working paper series, No 12, Mandaluyong City, Philippines, Asian Development Bank

OECD (2013) OECD secretary general Angel Gurria opening remarks to OECD forum people first! (28 May 2013)

Prahalad CK, Ramaswamy V (2000) Co-Opting customer experience. Harv Bus Rev 78(1):79–87

Ranieri R, Ramos RA (2013) Inclusive growth: building up a concept. In: International policy centre for inclusive growth working paper, p 104

Ravallion M, Chen S (2003) Measuring pro-poor growth. Econ Lett 78(1):93–99

Simanis E (2012) Reality check at the bottom of the pyramid. Harv Bus Rev. (June)

UN General Assembly (2005) The centrality of employment to poverty eradication: report of the secretary general August, A/60/314

UNDP (2010) The real wealth of nations: pathways to human development. Human development report

White H, Anderson E (2000) Growth versus distribution: does the pattern of growth matter?. Dev Policy Rev 19 (3): 267–289)

Chapter 6
The Value of 'Measuring' RRI Performance in Industry

Steven M. Flipse and Emad Yaghmaei

Abstract Within the EU funded project 'PRISMA' (Piloting Responsible research and innovation in Industry: a roadmap for tranSforMAtive technologies, 2016-2019) we aim to (Cooper and Kleinschmidt 1995) help industries implement Responsible Research and Innovation (RRI) strategies in their innovation processes, as part of their CSR policy, and (2) to provide evidence on how an RRI approach and explicit attention for the gender dimension can improve the innovation process and its outcomes. As part of our approach, we developed a method to intuitively score performance on certain RRI dimensions. The scoring method builds upon an earlier innovation quality monitoring tool, called the 'success factor based live innovation project scoring and evaluation' tool, and expands this tool with specific RRI elements. In this paper, we elaborate on the development of this scoring tool, and its potential value as a strategy to implement RRI in industrial innovation practice.

6.1 Introduction

Among some other things, the PRISMA projects aims for integration of RRI in the CSR (corporate social responsibility) policies of 8 pilot companies that use transformative technologies in the fields of biotechnology, nanotechnology, digital services and transportation. We intend to do so by providing evidence on how the RRI approach, and specific attention for the gender dimension, can improve the innovation process and its outcomes. Within the project, we develop a 'roadmap' that helps industries to implement RRI in their innovation processes in order to deal with uncertain and sometimes partly unknown risks and public and ethical concerns of transformative technologies. We hope that ultimately we can get a broad uptake of this roadmap by companies through industry associations, CSR and branch organizations, as well as governments and civil society organizations.

S. M. Flipse (✉) · E. Yaghmaei
Delft University of Technology, Delft, Netherlands
e-mail: S.M.Flipse@tudelft.nl

F. Ferri et al., *Governance and Sustainability of Responsible Research and Innovation Processes*, SpringerBriefs in Research and Innovation Governance,
https://doi.org/10.1007/978-3-319-73105-6_6

In our quest for relevant methods to help industry, we tried to find approaches that are specifically designed to help industry adopt RRI, not only in their CSR policies, but in actual, on-going R&D work as well. The RRI Toolkit can play an important role for our project, considering that the database of tools can possibly be deployed by our 8 pilot organizations, in dedicated industrial research and development (R&D) projects in which we will aim to explicitly integrate RRI.

Nevertheless, we aim to take this approach one step further: we also wish to find evidence of how RRI can help improve the innovation process and its outcomes. We found that in literature such evidence is still largely lacking, which begs the question: what is the value of RRI from the perspective of industrial innovation? Researchers from the social sciences and humanities have proposed that RRI can help the outcome of innovation processes in becoming more socially robust, and possibly, hence, more financially rewarding in a socially responsible way. This may or may not be the case, but that still begs the question what the value is of RRI practices in the 'midstream' (Fisher et al. 2006) of industrial innovation: on the R&D working floor.

In our project approach, we aim to do find such evidence (or the lack thereof) through both a qualitative and a quantitative assessment. Regarding the quantitative assessment, we will conduct interviews and use field observations to find leads that indicate that our RRI projects indeed perform differently from 'other' projects without specific attention to RRI using dedicated RRI approaches. However, the topic of this particular paper considers the *quantitative* approach that we also aim to develop. The remainder of this paper considers the setup of our approach and how we perceive of its possible functionality.

When thinking of RRI-relevant aspects of innovation on the actual, on-going R&D work by innovators working on transformative technologies, we feel we do have to start from the perspective of those who actually carry out the work on the laboratory working floor (Schuurbiers and Fisher 2009). That is where our methodological approach starts.

6.2 Methodology

Our overall methodology over the course of the PRISMA project consists of five methodological steps. Currently, we are working on step three of the process.

Step 1: Literature review—To include both obvious and less obvious RRI dimensions on the product and process level, we first wanted to find which indicators could theoretically and practically contribute to making such a process 'responsible', and then assess what the value is of scoring/ assessing these aspects (from whichever perspective). To find such indicators, we explored both literature in the field of innovation management (see Cooper and Kleinschmidt 1995, Van der Panne et al. 2003, Tepic et al. 2013, Flipse et al. 2013) and responsible innovation (see Hin et al 2014, Raven et al. 2015, Scholten et al. 2016).

From these papers and reports we distilled roughly 250 indicators. We reduced this number to a list of 92 indicators by removing redundancies and irrelevant indicators. We then reformulated all these indicators into statements about R&D processes, that people might agree or disagree with to a certain extent, in preparation of the later 'scoring' of these elements on a 7-point Likert scale. These remaining indicators we subsequently clustered collaboratively with the authors and external advisors into themes, relating to organizational R&D aspects and specific RRI criteria, both on the product and on the process level of innovation practice.

Step 2: Workshop for individual companies—In April 2017 we hosted a workshop for the companies participating in the PRISMA project. Six of the eight involved organizations were represented here (Email the corresponding author for the list of participants of the workshop), the remaining companies still need to contribute to this particular methodological step in the near future.

In the first step of this phase (*2a*), taking roughly 60–75 min, representatives of each company were asked to distribute our 92 indicators into three categories: absolutely relevant to their on-going R&D work, absolutely not relevant, and possibly relevant.

In the second step (*2b*), taking roughly 30–45 min, the indicators that were not considered relevant were discarded for each company while the remaining indicators were clustered into categories by the company representatives. This clustering is different than the clustering we did by ourselves (*Step 1c*), but this guarantees that the elements cluster into categories that are relevant for the individual companies.

In the third step (*2c*), taking roughly 15–30 min, the company representatives were asked to distribute a total of 100 points over their identified categories, to determine which categories they found to be the most important with regard to their influence on the quality of ongoing R&D work. The groups were free to determine their own approach towards the point distribution process (e.g., each person in a 4-person group gets 25 points, or 20 'rounds' of 5 points distribution per person, or any other approach).

Step 3: Tool development and first analysis—Considering we have not yet set up the tool for use by the companies, we cannot yet describe how we did *step 3a*. As such, we describe our plan here. The scoring method builds upon an earlier innovation quality monitoring tool, called the 'success factor based live innovation project scoring and evaluation' (sFLIPSE) tool (Flipse et al. 2015), and expands this tool with specific RRI elements. The tool basically asks for indicators to be entered, clustered into key performance indicators, and a score that determines the mathematical relative value of each of the clusters in relation to each other (*step 2c*).

With regard to our analysis (*step 3b*), we aim to determine initial similarities and differences between the participating companies. We therefore took our own clustering (*step 1c*) as a starting point, and identified to what extent the participating companies recognized the same individual indicators within these clusters. We also looked which indicators were always and never chosen, and which indicators were found to be unclear, or wrongly formulated. We also compared the clusters that the companies identified, to see if they recognize the same clusters of indicators.

Step 4: Use of tool in 8 pilot organizations—Our tool functionality requires RRI project participants (and possibly participants of other projects, within specific RRI efforts) to periodically fill in a questionnaire, consisting of the list of identified indicators (statements). We envision at least three monitoring points: at the beginning, midway, and end of a project's runtime.

Step 5: Meta-analysis—Analyses are mainly geared towards finding out if projects, in which specific RRI efforts were deployed, differ in performance or performance development from projects without such specific efforts. We will add here also the results of a subsequent qualitative analysis of our study, to identify reasons why RRI performance progressed the way it did. Herewith, we hope to conclude whether RRI efforts can support the quality of ongoing R&D work, and see whether from the perspective of the quality R&D process, it makes sense to explicitly adopt certain RRI methodologies.

6.3 Results

Step 1—In our literature study, we identified 92 different possible organizational and RRI indicators. RRI performance indicators and clusters provide an overview from literature in the field of innovation management and responsible innovation and do not constitute an evaluation of these items (MoRRI project (Metrics and Indicators of RRI) (Ravn et al. 2015) and a report on RRI indicators from European Commission 2015).

Step 2—In our workshop, we first asked industry representatives to identify, from the stack of 92 indicators, which indicators they considered 'relevant', 'irrelevant' or 'maybe relevant' for their organizations.

Out of 92 indicators, 89 functioned as a 'relevant' for companies in one or more situation. Of these, 11 indicators emerged as 'relevant' for all companies, while 7 indicators function as 'relevant' for 5 companies and marked as 'maybe' for the last company.

In order to help derive our organizational and RRI indicators for companies, companies have identified relevant indicators within their projects from 49 organizational and 43 RRI indicators. Companies show extensive variation in their selection of relevant indicators under organizational indicators and RRI counterparts. An overview is presented in Table 1.

We asked the companies to cluster the 'relevant' items into clusters of items. The comparison of clusters of indicators identified by six pilots demonstrates that there are clearly common clusters among companies:

Table 1 Categorized indicators by company for 6 of our involved companies

		Relevant	Irrelevant	Maybe	Relevant (%)
Company 1	Organizational	23	15	8	47
	RRI	24	15	4	56
Company 2	Organizational	36	7	4	73
	RRI	30	13	0	70
Company 3	Organizational	33	11	3	67
	RRI	21	14	8	49
Company 4	Organizational	36	11	0	73
	RRI	29	14	0	67
Company 5	Organizational	17	22	7	35
	RRI	27	12	4	63
Company 6	Organizational	24	9	14	49
	RRI	29	1	12	67
Total	Organizational				57
	RRI				62
	All				60

- Market and marketing related clusters
- Social and environmental related clusters
- Stakeholder related clusters
- Technology related clusters

We also asked the companies to score the relative importance of the identified clusters compared to one another, by distributing 100 points over their identified clusters (Email the corresponding author for a copy of results). Although the common clusters' importance may differ from one company to another, it was evidenced that these clusters were likely seen as important for other companies working with new technologies, albeit at varying rates.

Step 3—We checked how the indicators that companies selected, relate to our own clustering of indicators. Table 2 shows an overview.

6.4 Discussion and Conclusion

Within the PRISMA project, one of our targets is to provide evidence on how an RRI approach, and specific attention to the gender dimension, can improve the innovation process and its outcomes. Our approach towards this evidence is based on the workshop and tool we presented above. We hope that our scoring and analysis demonstrate an increase in performance of projects; i.e., both industrial projects with and without specific attention by using RRI methods. Yet maybe the RRI projects show a different performance development pattern, that can be

Table 2 Comparison of our framework to the indicators identified by our 6 companies

Organizational	# process/# product items	Co 1	Co 2	Co 3	Co 4	Co 5	Co 6
Internal							
Technology	5/4	3/3	4/3	4/3	4/2	2/1	4/3
Sales/marketing	2/3	2/3	2/2	2/3	2/3	1/2	0/3
Planning/management	4/1	2/1	3/1	2/1	4/1	0/1	1/0
Resources	4/1	1/1	4/1	2/1	3/1	1/0	0/1
Collaboration/ communication	2/0	0/0	0/0	0/0	0/0	0/0	0/0
External							
Market	0/8	0/3	0/6	0/7	0/6	0/2	0/4
Customer/end-user	5/10	1/3	1/9	0/9	4/6	0/7	0/8
RRI							
Diversity and inclusion							
Gender equality	2/1	1/1	0/0	1/0	2/1	1/0	2/1
Engagement	8/2	5/1	6/1	3/1	4/2	6/1	3/1
Anticipation and reflection							
Legislative landscape	2/1	1/0	2/0	0/1	1/1	2/0	0/1
Assessment	3/3	1/2	1/2	2/2	3/3	3/1	3/2
Public and ethical issues	1/1	0/0	0/1	0/1	0/0	1/0	1/1
Responsiveness and adaptive change	3/0	2/0	3/0	3/0	3/0	3/0	3/0
Openness and transparency							
Intellectual property and confidentiality	2/1	0/1	2/1	1/0	2/0	2/0	1/0
Open access	3/2	1/0	2/2	1/1	2/1	1/1	2/2
Environmental Sustainability	1/3	1/3	1/3	0/2	1/3	1/1	1/3
Social Sustainability	1/3	1/3	1/2	1/1	0/1	1/2	1/1

explained by circumstances that may or may not be related to RRI specifically. Therefore, it is still very important to also collect data in a qualitative manner, in addition to our intended scoring of projects at the beginning, middle and end of their runtime.

In any case, the assessment method makes us ask the following question: are we 'measuring' the initiative of the company itself, or of ourselves (or others)? We acknowledge that, similar to ethnographic methods, we cannot measure without influencing the practice we're studying. Even with this assessment method, the question remains whether and to what extent companies can organize RRI by themselves, and demonstrate that it works in practice. Therefore, within our project, we can explore the value of our approach and its potential value as a strategy to implement RRI in industrial innovation practice, but it will be difficult to continue doing such assessments when our project ends. Future studies could also consider what the value of RRI is without *explicitly* making it part of projects in industry, what options there are for independent RRI assessments and certifications, and maybe what options there are for integrated reporting on both RRI and annual statements/reports.

References

Cooper RG, Kleinschmidt EJ (1995) Benchmarking the firm's critical success factors in new product development. J Prod Innov Manag 12:374–391

European Commission (2015) Indicators for promoting and monitoring responsible research and innovation. DG Res Innov ISBN 978-92-79-43169-2 https://doi.org/10.2777/9742

Fisher E, Mahajan RL, Mitcham C (2006) Midstream modulation of technology: governance from within. Bull Sci Technol Soc 26(6):485–496

Flipse SM, Van der Sanden MCA, Van der Velden T, Fortuin FTJM, Omta SWF, Osseweijer P (2013) Identifying key performance indicators in food technology contract R&D. J Eng Technol Manag 30:72–94

Flipse SM, Van Dam KH, Stragier J, Oude Vrielink TJC, Van der Sanden MCA (2015) Operationalizing responsible research and innovation in industry through decision support in innovation practice. J Chain Netw Sci 15(2):135–146

Hin G, Daigney M, Haudebault D, Raskin K, Bouche Y, Pavie X, Carthy D (2014) Introduction to a guide to entrepreneurs and innovation support organizations. EU funded project report by Paris Region Enterprises and Knowledge Acceleration Responsible Innovation Meta (KARIM) network 1–60. (Nahuis R, Moors EHM, Smits REHM, (2012) User producer interaction in context. Technol Forecast Soc Chang 79(6):1121–1134)

Ravn T, Nielsen MW, Mejlgaard N (2015) Metrics and indicators of responsible research and innovation. Progress report D3.2 of the EU-funded Monitoring the Evolution and Benefits of Responsible Research and Innovation (MoRRI) project. 1–92

Scholten V, Cuppen E, Flipse S, Calon R, Van den Hoven J (2016) Rewarding RRI—a case study collection of the european foundations award for responsible research and innovation. A publication of the King Baudouin Foundation, pp 1–85

Schuurbiers D, Fisher E (2009) Lab-scale intervention. EMBO Rep 10(5):424–427

Tepic M, Kemp R, Omta O, Fortuin F (2013) Complexities in innovation management in companies from the European industry a path model of innovation project performance determinants. Eur J Innov Manag 16(4):517–550

Van der Panne G, Van Beers C, Kleinknecht A (2003) Success and failure of innovation: a literature review. Int J Innov Manag 7(3):1–30

Chapter 7
Responsible Research and Open Innovation in Geospatial Applications: Some Good Practices for Smart Communities

Fulvio Ananasso, Sergio Farruggia, Renzo Provedel and Monica Sebillo

Abstract The paper focuses on a specific case study within the Responsible Research and (Open) Innovation research field investigated by Stati Generali dell'Innovazione (SGI). It represents a work in progress specifically addressed to the benefit of human kind through the support of Geospatial applications. By promoting collaboration among scientists, societal actors, Institutions, the political sector and citizens, SGI, a not-for-profit association established in 2011, pursues the goal of facilitating a shared perspective for innovation policies in Italy that aligns with the Responsible Research and Innovation concept, as part of the Europe 2020 strategy (Ananasso et al. Responsible research and (Open) innovation to the benefit of human kind: the approach of Stati Generali dell'Innovazione, 2017). This paper describes SGI's initiatives aiming at improving awareness of Geographic Information value, thanks to the involvement of geo- competence and skill profiles related to (and requiring) openness, transparency and sharing of relevant science and innovation efforts and outcomes.

F. Ananasso (✉) · S. Farruggia · R. Provedel · M. Sebillo
Stati Generali dell'Innovazione, Rome, Italy
e-mail: fulvio.ananasso@alice.it

F. Ananasso
CDTI, Rome, Italy

S. Farruggia · M. Sebillo
AMFM GIS Italia, Rome, Italy

R. Provedel
FareImpresa, Genoa, Italy

M. Sebillo
University of Salerno, Fisciano, Italy

F. Ferri et al., *Governance and Sustainability of Responsible Research and Innovation Processes*, SpringerBriefs in Research and Innovation Governance,
https://doi.org/10.1007/978-3-319-73105-6_7

7.1 Introduction

Geospatial Revolution commonly refers to the transformation from analog to digital cartography, underpinned by Information and Communication Technologies (ICT) development in the framework of the digital information society. Indeed, the geospatial technology sector (Geo-ICT) has been undergoing an epochal change, overcoming the borders of traditional geomatics domain while involving many other ICT fields that use spatial information and maps, with a huge increase of new products and services, as well as the possibility to improve or re-engineer business processes (Geospatial Media and Communications 2017).

Data represents a basic proof of such an evolution. The increasing usage of geospatial data within new ICT applications, such as Big Data and Internet of Things (IoT), and the growing adoption of mobile devices by people worldwide, have been assigning to it more and more relevant roles and values. A study commissioned by Google to Oxera (2013) investigated direct, consumer and wider economic effects of the impact of geo-services (satellite receivers and manufacturing, electronic maps, satellite navigation, satellite imagery and location-based search) on the world economy and consumer welfare. Analysts estimated geo-services global revenues between US$150 billion and 270 billion per year. Such figures are higher than the US$25 billion revenues generated by the videogames industry, broadly equivalent to the US$140 billion revenues from the global security services industry and around one-third of the global airline industry revenues (US$594 billion).

Oxera study dates back to 2012. Nowadays, due to the rapid growth of the Geo-ICT industry, its conclusion can be considered an underestimation of the present and next future. The economy and society through the geospatial industry has been subsequently updated to over US$500 billion worldwide (Datta 2016).

This adjustment should be discussed also in terms of goals set by the UN Agenda 2030 for Sustainable Development. Indeed, the heritage of data collections with a spatial component is strongly related to the actualization of such an Agenda and its Sustainable Development Goals (SDG). In particular, that UN document emphasizes the need to adopt new ways for data acquisition and integration in order to improve their availability and quality for a more fruitful utilization in many domains including earth observation and geospatial information.

The first annual report by the UN Secretary-General (United Nations, 2016) on the Agenda 2030 reaffirms this concept and endorses the importance of establishing partnerships with the civil society, the private sector and the academic world. Moreover, it confirms the fundamental role of the integration between geospatial information and statistical data for the production of a number of indicators.

Stati Generali dell'Innovazione (SGI) is working to contribute to these goals. SGI, a social promotion association, was born on the initiative of some associations, movements, companies and citizens convinced that the best growth opportunities for Italy/EU are offered by young people's creativity, acknowledgment of merit, fighting digital divide, and State modernization through Open Government.

This paper describes some initiatives carried out by SGI, together with AM FM GIS Italy[1] (www.amfm.it), focused on topics essential for promoting, upgrading and actuating government innovation programs. Through an inclusive process, SGI aims to create awareness among citizens about the role they can have in decision-making processes. In order to achieve this goal, it is crucial to provide the general public with information, methods and tools useful to support their (daily) activities. In the following, some areas related to geographic information are described, where SGI and AM/FM GIS Italy are particularly active, i.e. sustainable development and education/training.

7.2 Geospatial Revolution Toward Open Innovation

Referring to the Open Government key principles (transparency, participation, collaboration) most leading Countries Governments have been adopting Open Innovation strategies and tackling the relevant associated issues (Lee et al. 2012).

Following the operating and social implications of the governmental open innovation policies, highlighted by the referenced study, SGI and AMFM GIS Italy are promoting, together with other stakeholders, novel professional profiles on the geospatial sector in order to address the impact of the Geospatial Revolution, taking into account 21st century paradigms, mainly those of Smart City/Community and Spatial Enablement (Farruggia and Sebillo 2016). In the present context, a smart city refers to a future city projection reflecting present cultural evolution and social perspectives, adhering to the Open Government concept. The latter identifies the ability to exploit geospatial data, both in terms of its availability and skills acquired for its utilization.

As a matter of fact, the definition of new professional profiles on the geospatial sector may allow smart cities/communities processes to become also "spatially enabled". The availability of innovative geospatial capabilities and competencies—for instance—allows public administrations to promote/utilize both the exploitation of their (geospatial) data, together with their needs, outside their own "walls" (inside-out open innovation approach) and the expansion of their knowledge base through the integration of external resources, such as citizens, SME and non-profit organizations—outside-in open innovation contributions.

Open innovation may represent a challenge *vis-à-vis* the huge amount of players involved in such a process—not only public bodies, regardless of relevance and size. The profiles of Geographic Professionals shown in the next section, aim at contributing on how governance bodies of smarts cities can use the huge potential

[1]Automated Mapping/ Facility Management GIS Italy and SGI have signed in 2015 a Memorandum regarding collaboration on innovative themes relevant to geographic infor- mation for citizen services, spatial/environmental data infrastructures, application/devel- opment of European Directives/Initiatives (e.g., INSPIRE and PSI Directive, the Digital Agenda), dissemination of good practices, training and certification of professional skills.

of Geo-ICT and geospatial data available within communities, following holistic approaches. Two examples: (i) the contribution on the matter within overarching strategic plans developed at smart city governance level; (ii) the support within communication processes and community-institution relationships regarding Open Innovation approaches joined with a territorial "prosumerism". In other words, whereas a territory provides users with information, the same territory may utilize information from users (User Generated Content, UGC), both activities being utilized to create synergies, effective data dissemination and planning strategies (Sebillo et al. 2015).

7.3 New GI Professional Profiles Taking Advantage of Open Innovation

In order to seize opportunities offered by Open Innovation paradigm within the digital geospatial revolution, it is crucial to face the geo-digital competencies issue, identifying professional profiles within the geospatial sector, such as Geographic Information (GI) managers capable of improving geospatial data usage within a Spatial Data Infrastructure (SDI),[2] and an enabler of spatial thinking within networks of stakeholders interested on building an intelligent community.

Stati Generali dell'Innovazione and AMFM GIS Italy are involved in several national initiatives dealing with geospatial sector. Specifically, such actions are based on their awareness that the actuation of Open Innovation requires the involvement of multidisciplinary competences, capable of both acquiring and using knowledge extracted from a territory/community, and integrating proper usage of available technology across evolving scenarios. In particular, SGI and AMFM GIS Italy are proactively taking part to an initiative by the UNINFO regulatory agency (2017) to establish a norm dealing with the regulation of geographic information profiles. The initiative, supported by AgID, the Governmental Authority responsible for implementing of the Italian Digital Agenda, aims at:

- identifying and characterizing the main professional profiles of the GI domain;
- establishing appropriate relationships with national and international regulators;
- defining GI skills, abilities and competences and integrating them with UNI 11506 ones[3];
- extending deliverables and KPI (Key Performance Indicator) to GI professional profiles in order to provide markets with definite reference and proper quality;

[2]Following the 2007/2/CE Directive (INSPIRE): "An SDI means: metadata, spatial data sets and spatial data services; network services and technologies; agreements on sharing".

[3]UNI 11056/2013: Unregulated Professional Activities - Professional figures in the ICT sector—Definition of knowledge, skills and competences requirements.

- identifying further improvements and refinements items;
- providing legislator with useful means to counterbalance the technical regulation and the mandatory field.

Preliminary results yielded so far two specific profiles, the Geographic Information Manager (GIM) and the Geographic Knowledge Enabler (GKE). GIM's main task is promoting the growth of quality level and technical expertise across the stakeholders' network. Her/his commitment is mainly focused on issues related to the capability of benefiting from spatial data (spatial enablement), both in terms of availability of such information and skills required for their utilization and valorization. GKE's main task is fostering the spatial awareness and thinking across the stakeholders' network. Her/his commitment is mainly focused upon issues related to geographic professional skills and "spatial citizenship" education, allowing individuals and groups to interact and participate in societal spatial decision making, through proper production and utilization of geo-media—geographic media such as maps, virtual globes, GIS, and the Geoweb. GKE must also monitor/update the required geospatial competencies, and promote proper education to grow competences on spatial thinking and geospatial intelligence, in order to provide future-proof, long lasting services and users spatially connected. GKE professionals may support GIM experts across complex organizational ecosystems.

7.4 Conclusion and Discussion

Geospatial sector offers development opportunities as never occurred in the past. Geo products and services may have a systemic impact on the triplet "economics-environment-society" (in other words, "sustainability") (Anansso et al. 2017).

The paradigm shift of geospatial information management, from a mere visualization and mapping tool to a fundamental, evidence-based decision-making solution, is inevitable and desirable. It may allow both the society and individuals to be aware of Geospatial Revolution impacts, being enabled to benefit, in a participative way, from the opportunities to improve citizens' lifestyle, quality of the environment, businesses and institutions, targeting a sustainable growth.

Within such a framework, a smart city/community needs to be "spatially-enabled" itself, i.e. (i) considering geospatial data as a common good and (ii) making them openly available in order to stimulate innovation (Roche et al. 2012). Then, three requirements have to be met to become a spatially-enabled smart city: (1) citizens must be able to enjoy geospatial data, also having the opportunity to acquire appropriate education, i.e. they need to be "spatially-literate"; (2) a favorable environment in place for sharing spatial data, adopting globally unified geospatial standards; (3) fair and smooth communication processes, whereby geospatial component is transparent and ubiquitous, so as to make the proliferation of community-institutions relationships fluid and effective.[4]

Innovation cannot be brought by technology only. GIM and GKE profiles respond to the needs of smart cities/communities willing to become also spatially enabled. Their competences and skills have been defined in order to allow them to play an important role in establish holistic approaches for managing the changes induced by the Geospatial Revolution, contributing to the diffusion of openness within their spatially enabled smart city/community. In this context, they will be able to promote both inside-out and outside-in open innovation approaches.

References

Ananasso F, Farruggia S, Provedel R, Sebillo M (2017) Responsible research and (Open) innovation to the benefit of human kind: the approach of Stati Generali dell'Innovazione (this issue)

Datta A (2016) Where is the money in geospatial industry? Geospatial World, 22 May 2016

Farruggia S, Sebillo M (2016) Smart cities and new professional opportunities: the geographic information manager, 19th AGILE International Geographic Information Science, June 2016, Helsinki, Finland. ISBN: 978-90-816960-6-7

Geospatial Media and Communications (2017) Global geospatial industry outlook January 2017

Lee SM, Hwang T, Choi D (2012) Open innovation in the public sector of leading countries. Manag Decis 50(1):147–162

Oxera (2013) What is the economic impact of Geo services?

Roche S, Nabian N, Kloeckl K, Ratti C (2012) Are 'SmartCities' smart enough? In: Global geospatial conference 2012, Québec City, Canada, 14–17 May 2012

Sebillo M, Tucci M, Tortora G, vitiello G, Ginige A, Di Giovanni P (2015) Combining personal diaries with territorial intelligence to empower diabetic patients. J. Vis Lang Comput Elsevier 29, 1–14. https://doi.org/10.1016/j.jvlc.2015.03.002

Uninfo Ente di Normazione Federato UNI (www.uninfo.it)

[4]The three requirements originally quoted by Roche et al. (2012), which a smart city has to meet in order to become spatially enabled, have been reworked by the authors of this paper.

Part II
Sustainability of RRI Processes

Chapter 8
Responsible Research and (Open) Innovation to the Benefit of Human Kind: The Approach of Stati Generali dell'Innovazione

Fulvio Ananasso, Sergio Farruggia, Renzo Provedel and Monica Sebillo

Abstract Economic growth, employment and sustainable development are amongst EU top priorities, relevant to a proper socio-economic (sustainable) development in the digital transformation society. Internet-based economic growth should then address targeted education (entrepreneurship, digital skills, specific competencies, etc.) as well as connecting research and innovation, bringing to the *Open Innovation* paradigm. A novel mentoring/Open Innovation initiative ("SGI Innovation Academy") under assessment at Stati Generali dell'Innovazione addresses from a Responsible Research and Innovation (RRI) point of view: (a) advanced education & training, (b) targeted RRI projects, and (c) support to new business initiatives/ startups, through coaching, mentoring, interim management, etc.

F. Ananasso (✉) · S. Farruggia · R. Provedel · M. Sebillo
Stati Generali dell'Innovazione, Rome, Italy
e-mail: fulvio.ananasso@alice.it

F. Ananasso
CDTI, Rome, Italy

S. Farruggia · M. Sebillo
AMFM GIS Italia, Rome, Italy

R. Provedel
FareImpresa, Genoa, Italy

M. Sebillo
University of Salerno, Fisciano, Italy

F. Ferri et al., *Governance and Sustainability of Responsible Research and Innovation Processes*, SpringerBriefs in Research and Innovation Governance,
https://doi.org/10.1007/978-3-319-73105-6_8

8.1 Stati Generali dell'Innovazione and Responsible Research and Innovation

Stati Generali dell'Innovazione (SGI) is a not-for-profit association established in 2011 with the goal of facilitating a shared perspective for innovation policies in Italy/EU (SGI Website 2018). It aims at:

- representing a benchmark for anyone involved in "innovation" at large, so as to promote synergistic approach of individual initiatives
- defining a "bottom up" path to organize on the territory policies for responsible innovation, through use of e-collaboration, thematic meetings, events, etc.
- elaborating/promoting programs for innovation in the Italy/EU governance.

In other words, SGI is deeply committed—in synergy with other associations, civil society representatives, political movements, institutions, etc.—to contribute improving EU Institutions and socio-cultural development of our Countries. Its goal is representing a think tank of "multipliers of social innovation", aiming at promoting a Country/EU Innovation Agenda.

SGI has been developing a unique, innovative working "format" (#scopriamolecarte, SGI Website 2018), implementing "innovation cards" (Innovation Cards 2017) from specific "innovation words". They show images and sentences stimulating critical reflection/thoughts on those specific topics, as a consequence of "coaching circles" generative dialogues, aiming at getting out of the *already known*, combining the principles and methods of Open Space Technology and "Theory U" (Scharmer 2016; MOOC MIT 2013). Each card is an access door (in augmented reality) to other images, feelings, thoughts from the minds and hearts of people. Once completed, "playing" with such a deck of cards in the framework of civic/institutional brainstorming and consultations would help stimulating generative dialogues/pro-actions on a shared perspective for Country/EU Innovation Agenda.

The Responsible Research and Innovation (RRI) concept, part of Europe 2020 strategy, has «the objective to promote a vision for a stronger collaboration among social, natural and physical scientists, societal actors and citizens in order to achieve a wider dimension of science and innovation and improve the role of society in environmental preservation» . The SGI scope of work is then totally in line with the RRI concept. We would add the consequent socio-economic development and welfare state preservation in a better harmonized and less unequal world.

Establishing and enlarging a RRI community with relevant stakeholders (scientists, societal actors, influencers, citizens,…) working together with the common goals of aligning RRI initiatives to values, needs and expectations of the society from science and innovation, has to do with SGI's efforts to improving Institutions, quality of life, younger generations welfare and development goals,…

The triplet "economics-environment-society" ("sustainability") gives the idea of a systemic/ethical vision we can apply as a metric to RRI projects, in terms of will and capacity to cope with substantial global challenges like (i) economic inequalities, (ii) environment deterioration and (iii) spiritual conflict between "ego-system" and "eco-system"—from our "past self" to the "future emerging self" (MOOC MIT 2013).

8.2 Jobs and Economic Growth in the Digital Transformation Society

Amongst the major social issues relevant to economic development, employment in the digital transformation society is one of the most serious emergencies. Substantially all types of jobs are affected one way or another by the digital transformation process, raising unemployment as a first effect. In the short-to-medium term, new information and communication technologies (ICT) inevitably cut jobs—not only less specialized ones. Then, specific ICT-related vocational training should hopefully open new opportunities in emerging sectors—Industry 4.0, Robotics, Artificial Intelligence, Data Science,… On top of that, platforms, algorithmic (e.g. Mechanical Turk) and gig economies are creating highly unstable, widely available and precarious job markets, providing working conditions at the boundary of unfairness, not contributing to people's welfare, pension, healthcare systems, etc.

However, even deeper education/specialization is not able to prevent unemployment to an intolerably high fraction of our youngsters—and no prospect of reversing such trend can be shortly seen on the horizon. By contrast, a majority of sector experts believe now that the gap between new jobs created and those destroyed by digitalization/internet economy will be negative, although others estimate, in the long run, five new jobs created every two jobs lost, extrapolating results from previous industrial "revolutions" and consequent turmoil prior to steady state adjustment. The whole matter deserves deeper assessment with regard to employment policies, the influence of such variables as education, digital literacy, technological development, demography and social inclusion, mobility, globalization, as well as lowering working time for achieving wider employment, sustainability of non-work-related income, free time management, and so on.

What does "responsible research and innovation" means in such a context, what kind of help can it give to us? For instance unveiling intangible needs and values where RRI can bring appropriate solutions, such as the "5 Rs" of sustainable logistics: **r**ecycling, **r**e-use, **r**eduction of scrapping/energy consumption, differentiated **r**ecycling. We need *jobs, products and services* matching such 5 Rs.

8.3 Research and Innovation for Entrepreneurship and Job Creation

It is then mandatory addressing and stimulating economic growth and development, enlarging the size of the goods and services production “pie”, and creating new (qualified) employment in the «digital economy» paradigm. Over 90% of new employment will require ICT knowledge and skills at various levels, and digital economy features and instruments should be utilized to the maximum possible extent in order to create new opportunities and jobs—in particular for young generations, preventing them from being “lost” generations or NEET (not in education, employment or training).

To this end, we should bear in mind the need for advanced education and meritocratic development/valuing of young talents (avoiding their diaspora, very costly in social and economic terms), and the mentioned transformations caused by digitalization and internet-related tools. To tackle the issue of digitalization-induced job cuttings, it follows the need to invest heavily (from school) on specific education—entrepreneurial and digital skills (way more profound than those of ‘digital natives’), critical thinking, systemic vision, leadership, etc.

Owing both to recurring economic crises and change of production paradigms in the digital economy, it becomes more and more difficult for young generations to find long term jobs. They should rather aim at developing business ideas and starting up their own endeavors. They should then be trained to the mentioned entrepreneurship and digital competencies since (at least) the undergraduate education, in order to prepare brilliant students to the knowledge society challenges, helping them to create business initiatives.

Thus, industrial development in the digital economy, promotion of young talents and new jobs creation require high specific education, internationalization of research, projects and innovation processes based on collaborative web platforms (e.g. crowdsourcing), identification of new businesses development opportunities and operating support to startup acceleration (coaching, mentoring, interim management, internationalization,…) and aggregation in order to reach critical mass.

In such a context, Universities and Research Centers would play a crucial role as far as the research ecosystem is concerned. In the framework of the digital economy, they should provide (a) high education on the knowledge society hot topics (including entrepreneurship and digital competencies, today little dealt with in academic education), (b) deeper web-driven support to job placement of graduate students and (c) internationalization of research, through participation to/ management of strategic cooperation projects and identification of new businesses/startups—targeting the best talents internationally.

Here comes the crucial distinction between *research* and *innovation*. A simple statement connecting them might be:

Research produces knowledge, *innovation* uses the knowledge to improve and generate products and services which are recognized as a value and paid by the market. No adoption by the market, would mean no innovation.

The term "value" has assumed recently a new semantic meaning, according to the *sustainability* paradigm. In turn, sustainability has been recognized as addressing *environmental, social* and *economic* aspects—the "triple issue". Hence, *innovation* should be checked and accepted should it pass a "sustainability test".

8.4 Open Innovation

Owing to the above, *innovation* is about products, services **and** *processes*—the *awareness of the process* is the key. Besides R&D people, new communities of contributors are increasingly involved in the business operation of enterprises side by side with employees—partners, startups, SMEs, solution providers (e.g. through crowdsourcing platforms), etc. These players in the "innovation framework" are called to contribute onto the innovation process thanks to the *open innovation* paradigm, rather than policymakers' decisions, in house R&D, etc.

Open Innovation has been a valuable and new paradigm since early 2000's (Chesbrough 2003a, b; Provedel 2007; Lee et al. 2012). It is a step ahead "innovation" because « open and courageous minds get solutions (even unexpected and unobvious) by a global eco-system » (of solvers worldwide). Open Innovation helps tackling crucial ("human driven" rather than "technology driven") challenges: (1) contrasting orthodoxy—e.g. (often flawed) believes "we already made that"; (2) utilization of available options—try and fail is better than conservatory status, experimenting novel combinations of factors; (3) intercept needs/pent up demand—unsolved problems/unsatisfactory solutions allow better definition of needs; (4) extrapolate trends/detection of "weak signals"—often in unexpected cross sectors not initially addressed ("serendipity")—may originate great opportunities.

Open Innovation greatly benefits from the mentioned "Theory U" (Scharmer 2016), requiring *open mind, open heart, open will* to discover the emerging future and side spots to launch actions. Linking "listening" with "prototyping" in every person leads to a new kind of leadership, the "emerging future leadership". At SGI, we are using this new paradigm to improve the connection mood among people and produce better ideas, knowledge and projects.

A major paradigm shift in innovation sees small organizations playing increasing roles, whilst large ones keep reducing in-house R&D by applying new strategies such as SME/startups acquisition, broad ecosystem of partners to jointly develop new solutions/technologies, etc.

8.5 MENTOR/SGI Innovation Academy ("InnovA")

As pointed out, *employment, economic growth* and *sustainable development* represent crucial topics in the digital transformation society, and thus should be high priority items on the agenda of a Responsible Research and Innovation framework. However, they would require a "friendly ecosystem", providing young people with entrepreneurial education and tools in the digital economy paradigm, promoting more concrete technology transfer from public research and open innovation features and tools (e.g. crowdsourcing). In such context and with those objectives, the SGI "Innovation Academy" aims at implementing a management team interfacing Education, Research, Innovation and Enterprises—in line with the EU paradigms and EIT "Knowledge Triangle"–, helping young people with (Ananasso et al. 2016):

1. advanced education (entrepreneurship, digital & specific competencies,…) by advanced tools like blended/adaptive learning, OER, etc.—e.g. platforms EdX, Coursera, Future Learn, OpenupEd, Federica,… providing small private/ distributed/massive open online courses (SPOC/DOCC/MOOC).
2. support and co-ordination of targeted research and (social) innovation projects, identification of new business initiatives/startups, etc., exploiting both spinoffs from research and open innovation approaches—such as Open health, Open Science, Geospatial revolution dealt with in separate papers.
3. support to spinoff/startup financing (crowdfunding, private equity,…)
4. coaching, mentoring, interim management, internationalization,…

Senior experts would beneficially assist young entrepreneurs on startup strategic/ operating issues, helping them on needs assessment, problem solving,, international business development/joint ventures (owing to their senior contacts), transfer of relevant know how,…, so as to accelerate development prospects.

Contamination Laboratories (C.Labs) are designed to expose students of different disciplines to entrepreneurship by stimulating inter-/multi-disciplinarity, learning places where skills and mindsets develop to transforming creative ideas into entrepreneurial activities. C.Labs pursue social economy objectives, promoting business ideas based on *civic responsibility, diligence* and *social solidarity*. Exposure to the *corporate social responsibility* (CSR) approaches facilitates training to systemic thinking and leadership.

SGI aims at linking C.Labs with initiatives of enhancement and smart management of community-based assets, that can deliver economic development and concrete opportunities of (new) employment in our territories under the RRI paradigm. We are also envisaging methods and approaches matching our Mediterranean origin, such as the Science/Arts/Technology/Society models to deal with issues such as migration, climate change, energy, data openness, inclusion

8.6 Conclusions—Discussion

The Responsible Research and Innovation approach of Stati Generali dell' Innovazione is matching the EU strategy of promoting collaboration among scientists, societal actors, institutions, the political sector, citizens,… in order to achieve a wider dimension of science and innovation, improve the role of society in environmental/welfare state preservation, its socio-economic development and less unequal world.

To this end, we have been addressing such relevant issues as **economic growth, employment** and **sustainable development**—from the environmental, social and economic point of view.

SGI Innovation Academy ("InnovA") aims at implementing a whole ("one stop shopping") coverage of the innovation value chain—*from Education to Market*—through a multi-disciplinary network of senior experts providing advanced education, advisory to R&I projects, support to spinoff/startup/SME funding, acceleration/interim management of new business initiatives.

Acknowledgements The authors greatly appreciate comments and suggestions by Salvatore Iaconesi.

References

Ananasso A, Ananasso F (2016) Infrastrutturazione e Crescita nell'Economia Digitale. INFORAV "Informatica & Documentazione" n.1/2016 (April 2016)

Chesbrough HW (2003a) The era of open innovation. MIT Sloan Manag Rev

Chesbrough HW (2003b) Open innovation: the new imperative for creating and profiting from technology. HBS Press. ISBN 978-1-4221-0283-1

Innovation Cards (2017) (#scopriamolecarte) https://www.statigeneralinnovazione.it/online/wp-content/uploads/2017/06/Carte-Innovazione-300-DPI.zip)

Lee S, Hwang T, Choi D (2012) Open innovation in the public sector of leading countries

MOOC MIT #15.671 (2013) Leading from the emerging future (U.Lab)

Provedel R (2007) L'innovazione aperta: come catturare le idee dal mondo attraverso i broker dell'innovazione. Mondo digitale, n. 1, marzo

Scharmer CO (2016) Theory U: leading from the future as it Emerges. Berrett-Koehler Publishers

SGI Website (2018) https://www.statigeneralinnovazione.it/online/

Chapter 9
RRI Approach for Development and Acceptance of Novel Fish Feed Formulations in Aquaculture

Chiara Magoni, Ausilia Campanaro, Andrea Galimberti, Chiara Pesciaroli, Stefano Bertacchi, Paola Branduardi and Massimo Labra

Abstract Stakeholders' involvement and public engagement in research processes are essential elements to obtain concrete impacts in society. In this context we tested the Responsible Research and Innovation (RRI) strategy in the aquaculture industry to improve fish quality by using the most suitable feeds. These could be obtained through biotechnological approaches starting from industrial waste, or by using flour with high nutritional values such as insect-meals. Both strategies were compared at the scientific level, to estimate their performance in terms of fish quality and health, and at the RRI impact. In this contest a pool of stakeholders was engaged to evaluate the economic, environmental and social aspects of both technological approaches and to steer research and technology transfer activities.

C. Magoni (✉) · A. Campanaro · A. Galimberti · C. Pesciaroli · S. Bertacchi
P. Branduardi · M. Labra
University of Milano-Bicocca, Milan, Italy
e-mail: chiara.magoni@unimib.it

A. Campanaro
e-mail: ausilia.campanaro@unimib.it

A. Galimberti
e-mail: andrea.galimberti@unimib.it

C. Pesciaroli
e-mail: chiara.pesciaroli@unimib.it

S. Bertacchi
e-mail: s.bertacchi@campus.unimib.it

P. Branduardi
e-mail: paola.branduardi@unimib.it

M. Labra
e-mail: massimo.labra@unimib.it

F. Ferri et al., *Governance and Sustainability of Responsible Research and Innovation Processes*, SpringerBriefs in Research and Innovation Governance,
https://doi.org/10.1007/978-3-319-73105-6_9

9.1 Problem Identification

Aquatic resources, with a total production of 167 million tons (FAO 2016), cover a significant percentage of the world proteins demand and almost completely the need for the rare Long Chain Fatty Acids omega-3, known to be essential in human nutrition (Evans and Burr 1927). In the last two decades, fishery has been overexploited, and it no longer maintained the same worldwide consumption standard of 20.1 kg per capita (FAO 2016). Therefore, since only aquaculture can supply the growing fish demand, fish feed sector should be expanded in a sustainable way, including the adoption of more sustainable feed formulations. Generally, fish feed is composed of a mixture of fishmeal and fish oil. However, since their demand and costs are increasing, producers are substituting them with plants flour and other ingredients. As a consequence, feed formulations are deficient in several essential elements such as amino acids and show palatability problems. Over the years, many attempts have been made to find alternatives to traditional fish feed and recently waste biomass was evaluated as possible source of feed ingredients. This alternative source requires technical, economic, social and environmental evaluations through the RRI strategy to guarantee the innovation for aquaculture industry.

9.2 Analysis of Scientific Solutions and RRI Strategy for Biotechnological Process

In order to respond to the growing demand for alternative fish feed, a sustainable, viable, non-GMO biotransformation producing the key lipid ingredients (omega-3) through the transformation of the by-product glycerol was proposed. These ingredients will partially replace the components of current fish feed permitting to avoid, at the same time, the use of fish oil for intake of (ω-3). This is the main goal of MYSUSHI project (Microalgae and Yeasts SUStainable fermentation for HIgh quality fish feed formulation, Cariplo Foundation, N°2015–0395) developed by a team of University of Milano-Bicocca and University of Insubria (Italy). Biotechnologies are pioneering tools for bringing innovation to society but, due to difficulties in the assess of economic feasibility and limits imposed by socio-ethical acceptance of novelties, scientific innovations are penalized in marketing. MYSUSHI project exploited the concept of RRI (Fisher et al. 2006) aiming at modeling the research process on industry expectations together with stakeholders of aquaculture industry. Some of the strategies, outcomes, and considerations on the use and success of RRI analysis applied to an aquaculture case study are presented.

9.3 Strategy and Results

In our activities, we consider the three phases of the RRI: Engagement phase, Communication phase and Technology Transfer phase. We focused our attention on the first two phases, which are based on interactions with stakeholders to define the most suitable technology transfer actions. Experiences integrating societal considerations into an academic research laboratory setting suggest that it may help research planning and stimulate creativity (Von Schomberg 2013). Therefore, our aim was to consider all aquaculture stakeholders feedback in order to shape the research in a better way and to plan the technology transfer.

9.3.1 *Identification and Engagement Phase*

In this first phase, we considered the largest freshwater fish producers in Europe. Aquaculture industry of Northern Italy was particularly investigated since it is one of the major producers of EU for trout. Therefore, we evaluated the trout industry stakeholders through bibliographic analysis, direct farms visiting and trade fair events. As a result of this extended analysis, three stakeholder classes were identified: industry actors, scientific community, and civil society (Table 9.1).

Table 9.1 Different types and numbers of engaged stakeholders are indicated in the table

Stakeholder classes		
Industry actors	Scientific community	Civil society
Aquaculture companies: 10 Feed producers: 5 Companies: 2 Intermediates: 7 Retailers: 1 Industries magazine: 1	Ecologists: 12 Fish physiologist: 13 Veterinaries: 24 Fish pathologists: 9	Students: 6 Citizens: 18 Category associations: 4

9.3.2 *Communication Phase*

In order to critically evaluate the scientific solution proposed by MYSUSHI, we considered different RRI tools to interact with stakeholders: dedicated surveys, one-to-one meetings and web communication. Specific surveys were designed aiming to tackle three kinds of issues: socio-ethical, technical-scientific and economic. These surveys were firstly proposed during Aquafarm fair to reach industrial stakeholders (Pordenone, 26–27 January 2017). This fair represents the most suitable environment for engagement phase. Results are described in Fig. 9.1. Data suggested that most people were highly interested in the optimization of fish feed

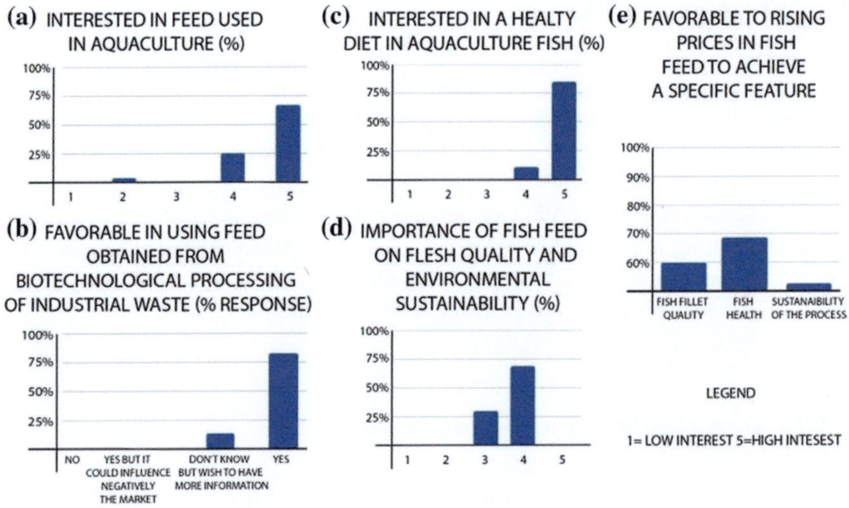

Fig. 9.1 Results of surveys' question. The stakeholder interest was expressed from 1 to 5 score. In **b** and **e** the specific features were indicated in the x-axis

(Fig. 9.1a) to enhance fish nutrition quality (Fig. 9.1c). The majority considered biotech processes a valuable tool to increase fish feed quality (Fig. 9.1b). Moreover, the environmental aspect represents a challenging issue of the aquaculture supply chain (Fig. 9.1d). Finally, almost a 70% of interviewed stakeholders were willing to pay for higher quality of fish fillet (Fig. 9.1e).

Industry actors were also engaged through one-to-one meetings to discuss all scientific project phases. Moreover, dedicated visits to the production facilities allowed us to better understand the organization of overall industrial process, guaranteeing the integration of research and technological goals. All tools adopted with industrial actors were very effective to improve research plan actions, to highlight the environmental impact of new feed formulations, to support bio-transformation in aquaculture and to enhance the fish quality aspects.

We planned the project considering local and European laws to design an innovation that could be implemented in the market. With more details about R&D innovation it will be possible to define the strategy with regulators. Among the stakeholders, scientists were reached through a strategy adopted by the grants office of Cariplo Foundation during MYSUSHI project evaluation. Three anonymous referees produced a technical report useful to improve scientific issues within the project, i.e. to identify a plan for biotechnological waste valorization phase. Moreover, one-to-one meetings were used specifically to evaluate the initial phase of the project during the selection of the most suitable strategy that could be used to improve fish feed formulation. In addition, MYSUSHI research team described the project during national and international conference through oral presentation and poster section (Aquafarm 2017, ASPA 2017). RRI action with scientific community

Table 9.2 Summary of possible feed alternative strategies, evaluated with scientific stakeholders

Source	Advantages	Disadvantages
Plant based products	• High availability • Good protein content	• Palatability and pathological problems • High presence of fiber stimulate more excretion • Lipid fraction rich in ω-6 instead of ω-3 • Impact on use of soil
Animal based products	• Use of by-products • Amino acid profile more complete than vegetable one • Lower price than fish protein	• Risk of disease transmission (TSE) • Need of mixing with polyunsaturated fats to avoid problem of fish digestibility • Consumer acceptance
GMO Trout	• Increased size of trout fillet (Medeiros et al. 2009)	• In Europe, use of GM animals is prohibited • Public acceptance of GM trout is hard to achieve
Insect based products	• Very fast production rate • Legalized use of insect flour	• Lipid content not sufficiently rich in omega-3 • Presence of chitin can interfere

allowed to evaluate alternative solutions of waste to produce feed as well as some scientist contributed to improve the nutritional quality of new feeds as described in Table 9.2. Civil society was reached through two different tools: a project dedicated website and social networks such as Facebook and Twitter. Facebook page "MySushi Biotech" has a total coverage of more than 9000 views while Twitter account (@MySushiBiotech) reached more than 38.000 views. We observed that citizens prefer to use Facebook to discuss about the aim of project while experts on specific issues (i.e. food systems) used Twitter to communicate with the project team. Those tools are also useful in order to contact non-commercial organizations since they provide full, yet simple explanation of the scientific solution and interaction with scientists. Initially, these tools were used to inform a wide public about MYSUSHI goals and strategy. Currently we are focusing on the presentation of critical issues of the project (ethical, social and economic) in order to stimulate a constructive debate to select the most active citizen-stakeholders.

9.4 Outcomes and Strategy for Next Phases

Our RRI analysis allowed to better identify the most suitable tools and strategies to reach several stakeholder classes. In the case of industrial actors we found one-to-one meeting the most promising strategy to drive research activities. However, considering that most of the industrial actors have a strong technical expertise, RRI team should involve not only experts in communication but also technical and scientific experts of the aquaculture sector. This could improve the relationships between industrial actors and scientists. Dedicated visits to the industrial facilities help the researchers in the identification of the main industrial criticisms, for example on the strategy for handling waste material for biotransformation. Concerning scientific stakeholders, we considered the classical ways of scientific debate (meetings, conferences, papers) the most suitable tool to improve project's quality. Finally, citizens represent the most critical class of stakeholders due to their heterogeneity and lack of information about the criticism of aquaculture sector. On the whole, the RRI strategy adopted in our preliminary phase was suitable only for the communication phase but new tools should be identified to involve the citizen in the definition of innovation strategies, for example by defining a dedicated course for the more interested stakeholders, and to improve the role of citizen associations in the RRI evaluation of environmental impact, cost of production and future development.

References

FAO (2016) The state of world fisheries and aquaculture 2016. Contributing to food security and nutrition for all, Rome, 200 p

Evans HM, Burr GO (1927) A new dietary deficiency with highly purified diets. Proc Soc Exp Biol Med 24(8):740–743

Fisher E, Mahajan RL, Mitcham C (2006) Midstream modulation of technology: governance from within. Bull Sci Technol Soc 26(6):485–496

Von Schomberg R (2013) A vision of responsible research and innovation. In: Owen R, Bessant J, Heintz M (eds) Responsible innovation: managing the responsible emergence of science and innovation in society. Wiley, pp 51–74

Medeiros EF, Phelps MP, Fuentes FD, Bradley TM (2009) Overexpression of follistatin in trout stimulates increased muscling. Am J Physiol Regul Integr Comp Physiol 297(1):R235–R242

Chapter 10
The MARINA Project: Promoting Responsible Research and Innovation to Meet Marine Challenges

Fernando Ferri, Noemi Biancone, Chiara Bicchielli, Maria Chiara Caschera, Alessia D'Andrea, Arianna D'Ulizia, Patrizia Grifoni, Tiziana Guzzo, Ned Dwyer, Tiago Garcia, Iwona Gin, Xenia Schneider, Elena Giusta, Saša Raicevich, Francesca Ronchi, Marco Picone, Cecilia Silvestri, Chiara Bolognini, Margot Bezzi, Chiara Buongiovanni, Emanuela Danè, Mara Gualandi, Nagore Ibarra Gonzalez, Mato Knez, Manuel Cira, Eoin Nicholson, Mihaela Mirea Candea, Angelica Paiu, Anca Gheorghe, Hans Thor Andersen, Jesper Rohr Hansen, Yiannis Laouris, Elena Aristodemou, Husne Altiok, Dilek Ediger, Helin Haga and Mathis Bogens

Abstract The MARINA project was funded by the European Union's Horizon 2020 Research and Innovation Programme with the aim to create an all-inclusive Knowledge Sharing Platform for catalysing and federating the convergence of already existing networks, communities, online platforms and services by

---

F. Ferri (✉) · N. Biancone · C. Bicchielli · M. C. Caschera · A. D'Andrea · A. D'Ulizia · P. Grifoni · T. Guzzo
IRPPS—CNR, 00185 Via Palestro, Rome, Italy
e-mail: fernando.ferri@irpps.cnr.it

N. Dwyer · T. Garcia
EurOcean, Avenida Dom Carlos I 126-2, 1249-074 Lisboa, Portugal

I. Gin
Nausicaà, Boulevard Sainte Beuve B.P.189, 62203 Boulogne-sur-Mer, France

X. Schneider
XPro Consulting, Nelson Street 7, 2021 Strovolos, Nicosia, Cyprus

S. Raicevich · F. Ronchi
Italian National Institute for Environmental Protection and Research (ISPRA), Loc. Brondolo, 30015 Chioggia, Italy

E. Giusta · M. Picone · C. Silvestri · C. Bolognini
Italian National Institute for Environmental Protection and Research (ISPRA), Via Brancati 48, 00144 Rome, Italy

M. Bezzi · C. Buongiovanni · E. Danè · M. Gualandi
APRE, Via Cavour 71, 00184 Rome, Italy

F. Ferri et al., *Governance and Sustainability of Responsible Research and Innovation Processes*, SpringerBriefs in Research and Innovation Governance,
https://doi.org/10.1007/978-3-319-73105-6_10

addressing marine research issues and topics according to Responsible Research and Innovation (RRI) principles. Several marine research issues and topics have been discussed in the 17 Mobilisation and Mutual Learning work-shops during the first year of the project. This paper shows the results of the first series of workshops analysed according to RRI principles and societal and marine challenges.

10.1 Introduction

The MARINA project[1] aims to federate Responsible Research and Innovation (RRI) communities and stakeholders working in the marine sector. This task is complex due to the heterogeneity of the stakeholders, the different agendas and priorities, the lack of communication among the different actors and the multiplicity of perspectives, interests and points of view in the current and future research and innovation activities. This complexity can be dealt with by a socio-technical approach (Dearden and Rizvi 2009; Coakes and Coakes 2005) that "refers to the interrelatedness of the 'social' and 'technical' aspects of an organisation or social structure" (https://russell.wiki.otago.ac.nz/index.php/SocioTechnical_Approac_-_STS).

The socio-technical theory establishes that any system and organization is characterised by a social sub-system and a technical sub-system. This perspective seems very useful for approaching the Responsible Research and Innovation (RRI) challenges

N. I. Gonzalez · M. Knez
CIC nanoGUNE, Tolosa Hiribidea 76, E-20018 Donostia–San Sebastian, Spain

M. Cira
ROM-WON, Louiselaan 89, Floor 7, 1050 Brussels, Belgium

E. Nicholson
SmartBay Ireland, Rinville, Oranmore, Galway, Ireland

M. M. Candea · A. Paiu · A. Gheorghe
Mare Nostrum, Bd. 1 Decembrie, 1918 3, 900711 Constanţa, Romania

H. T. Andersen · J. R. Hansen
AAU, Fredrik Bajers Vej 5, 9220 Aalborg, Denmark

Y. Laouris · E. Aristodemou
CNTI, Promitheos 5, Nicosia, Cyprus

H. Altiok · D. Ediger
İstanbul Üniversitesi Merkez Kampüsü, Beyazıt/Fatih, 34452 İstanbul, Turkey

H. Haga · M. Bogens
Science Centre AHHAA Foundation, Sadama 1, 51004 Tartu, Estonia

[1]The project "Marine Knowledge Sharing Platform for Federating Responsible Research and Innovation Communities—MARINA is funded by the European Union's Horizon 2020 Research and Innovation Programme under GA No. 710566.

(Regulation (EU) No 1291). The concept of Responsibility emphasises ethical issues at a collective level and attributes a prospective responsibility to a heterogeneous group of actors including States, the scientific community, industry and civil society. The aim is to build a common understanding of how to assess, appraise, manage and communicate risks that science is not yet able to fully evaluate "anticipating potential environmental, health and safety impacts and taking due precautions, proportional to the level of protection, while encouraging progress for the benefit of society and the environment" (Precautionary principle).

10.2 The Socio-Technical Approach

The socio-technical thinking facilitates to address multifaceted problems with the cooperation of all the actors combining the use of powerful information and communication technologies with the individual and collective multifaceted peoples' experiences and knowledge as a unique ecosystem (Grifoni et al. 2014; Ferri et al. 2013, 2014). Building a consensus for marine issues must take into account the various societal issues related to environmental conservation and the development of productive activities related to the sea (blue growth), with resulting impacts on people's lives and society itself. The MARINA socio-technical approach is built upon the engagement of policy makers, the scientific community, industry and civil society through Mobilization and Mutual Learning (MML) workshops, participatory methodologies, knowledge sharing and a community established in the Knowledge Sharing Platform (KSP). This Platform helps to bring together existing networks, communities, online platforms and services. Furthermore, the KSP allows sharing lessons learned from MML workshops and knowledge arising from other projects, identifying best practices on RRI, developing policy recommendations and sharing the outcomes in a travelling exhibition.

This paper presents the first results of 17 local MML workshops carried out in 12 European countries during the first year of the project, the knowledge outputs and lessons learned, all shared on the KSP. These MML workshops tackle, at the same time, a range of societal challenges and the RRI dimensions[2] at a pan-European scale by focusing on the marine domain as the practical thematic context for the implementation of its activities. The adopted socio-technical model consists of the following two levels:

1. *the social subsystem* that involves actors and communities in participatory, sharing and co-creation activities for acquiring and producing desired outcomes. Among these activities, the MML workshops enable the integration of contributions by individuals with diverse views, backgrounds, and perspectives. In an attempt to federate communities and raise awareness of the RRI, the 17 MML workshops engaged citizens, researchers, business representatives and policy makers (i.e. the RRI actors), in adopting the RRI approach in the marine sector

[2]Governance, Ethics, Public Engagement, Science Education, Open Access and Gender.

through identifying actions and solutions towards current and emerging marine and societal challenges;
2. *the technical subsystem* consisting of the Knowledge Sharing Platform (KSP) that contains knowledge outputs arising from synergies and interactions among the stakeholders (www.marinaproject.eu).

10.3 First Results of the MARINA Socio-Technical Approach

The 17 MML workshops involved 402 stakeholders (81 citizens, 66 local, national and international policy makers, 65 business representatives, 104 scientists, 58 representatives of NGOs and CSOs, 24 students and 4 journalists) and discussed marine issues and societal challenges based on local interests and priorities. Six workshops discussed sustainable tourism and coastal cities, five workshops focused on pollution caused by human, land and sea pressures, three workshops concentrated on fishing and aquaculture, two were dedicated to renewable energy and one was about marine changes caused by climate. The technical subsystem (KSP) supported the workshops with an open online debate by providing: (i) possibilities to share different kinds of documents, (ii) chats, (iii) videos, (iv) and online streaming, thus enabling wide participation before, during and after the event. During each workshop, the participants reached a common understanding of the discussed issues, identified a set of objectives and actions that were classified according to the related H2020 Societal Challenges[3] and the RRI dimensions. The expertise of the stakeholders involved in the workshops has contributed to populating the KSP with knowledge, building common understanding and sharing concepts emerged from the debates.

The results of the workshops have demonstrated that the marine sector is very pervasive in terms of societal challenges defined by the H2020 strategy for Europe. In spite of a limited number of topics discussed in the workshops, each H2020 societal challenge is implied in the outcomes (objectives/actions) in at least one MML workshop. The most prevalent societal challenges (ranked by the number of proposed actions) are "Climate action, environment, resource efficiency and raw materials"; "Europe in a changing world—inclusive, innovative and reflective societies", and "Food security, sustainable agriculture and forestry, marine and maritime and inland water research, and the Bio-economy". Table 10.1 shows the percentages of actions per workshop and per Societal Challenge.

[3]The H2020 Societal Challenges are: Health, demographic change and wellbeing; Food security, sustainable agriculture and forestry, marine and maritime and inland water research, and the Bioeconomy; Secure, clean and efficient energy; Smart, green and integrated transport; Climate action, environment, resource efficiency and raw materials; Europe in a changing world—inclusive, innovative and reflective societies; Secure societies—protecting freedom and security of Europe and its citizens.

Table 10.1 Percentage of actions per workshop and societal challenge

MML workshops		Societal challenges						
		Health, demographic change and wellbeing (%)	Food security, sustainable agriculture and forestry, marine and maritime and inland water research, and the Bio-economy (%)	Secure, clean and efficient energy (%)	Smart, green and integrated transport (%)	Climate action, environment, resource efficiency and raw materials (%)	Europe in a changing world—inclusive, innovative and reflective societies (%)	Secure societies—protecting freedom and security of Europe and its citizens (%)
Tourism and coastal cities	Tourism provides income but also destroys the values that brought it to Cyprus—How can we preserve both?	0,00	11,11	0,00	0,00	50,00	38,89	0,00
	Live the sea with those who live off the sea	1,87	6,54	0,00	17,76	42,99	30,84	0,00
	Marine protection in the shadow of urbanisation: prince's Islands case	0,00	33,33	0,00	0,00	38,89	27,78	0,00
	Sun, sand… and safety. How long will we have beaches?	0,00	12,82	0,00	0,00	20,51	46,15	20,51
	Connecting harbour and city—strategy, collaboration and growth	0,00	0,00	0,00	0,00	60,00	40,00	0,00
	Crowded city at the pristine coast	0,00	0,00	0,00	0,00	63,16	36,84	0,00

(continued)

Table 10.1 (continued)

MML workshops		Societal challenges						
		Health, demographic change and wellbeing (%)	Food security, sustainable agriculture and forestry, marine and maritime and inland water research, and the Bio-economy (%)	Secure, clean and efficient energy (%)	Smart, green and integrated transport (%)	Climate action, environment, resource efficiency and raw materials (%)	Europe in a changing world—inclusive, innovative and reflective societies (%)	Secure societies—protecting freedom and security of Europe and its citizens (%)
Pollution caused by human land and sea pressures	Marine litter: from the land to the sea and then back to us in our seafood?	0,00	45,12	0,00	0,00	35,37	19,51	0,00
	Pollution by human pressure with a special focus on marine litter	0,00	22,00%	0,00	0,00	52,00	26,00	0,00
	Water sustainability	0,00	0,00	0,00	0,00	46,81	53,19	0,00
	Marine litter and plastic pollution: toward a comprehensive and inclusive set of solutions	0,0	4,55	0,00	0,00	40,91	54,55	0,00
Sustainable fishing and aquaculture	Sustainable aquaculture—what is at stake for the future?	1,32	37,50	0,00	0,00	19,74	41,45	0,00
	Will Italian fish products disappear from our dishes?	3,45	67,24	0,00	0,00	0,00	29,31	0,00
	Responsible research and innovation for sustainable	0,00	1,96	0,00	1,96	31,37	64,71	0,00

(continued)

Table 10.1 (continued)

MML workshops		Societal challenges						
		Health, demographic change and wellbeing (%)	Food security, sustainable agriculture and forestry, marine and maritime and inland water research, and the Bio-economy (%)	Secure, clean and efficient energy (%)	Smart, green and integrated transport (%)	Climate action, environment, resource efficiency and raw materials (%)	Europe in a changing world—inclusive, innovative and reflective societies (%)	Secure societies—protecting freedom and security of Europe and its citizens (%)
	fishery and aquaculture: opportunities and synergies between the fisheries and aquaculture sectors in Azores							
Renewable energy	Wave energy in Ireland: considering the future—perspectives on policy	0,00	2,13	51,06	2,13	17,02	27,66	0,00
	Wave energy in Ireland: Considering the future—an industry perspective	0,00	10,53	76,32	2,63	0,00	10,53	0,00
Marine changes caused by climate	To adapt and survive or to die. How to maintain and enhance the quality of the Black Sea NW coastal ecosystems under the pressures of climate change and human interventions	0,00	46,08	0,00	0,00	53,92	0,00	0,00
	Average	0,39	17,70	7,49	1,44	39,57	32,20	1,21

Table 10.2 Percentage of actions per workshop and RRI dimension

MML workshops		RRI dimension					
		Governance (%)	Ethics (%)	Public engagement (%)	Science education (%)	Open access (%)	Gender (%)
Tourism and coastal cities	Tourism provides income but also destroys the values that brought it to Cyprus—How can we preserve both?	18,75	31,25	18,75	18,75	12,50	0,00
	Live the sea with those who live off the sea	46,25	7,50	38,75	6,25	1,25	0,00
	Marine Protection in the Shadow of Urbanisation: Prince's Islands Case	30,95	11,90	14,29	23,81	19,05	0,00
	Sun, sand… and safety. How long will we have beaches?	32,93	12,20	21,95	17,07	15,85	0,00
	Connecting harbour and city—strategy, collaboration and growth	47,83	21,74	30,43	0,00	0,00	0,00
	Crowded city at the pristine coast	27,27	0,00	33,33	36,36	3,03	0,00
Pollution caused by human land and sea pressures	Marine litter: from the land to the sea and then back to us in our seafood?	64,94	2,60	19,48	9,09	3,90	0,00
	Pollution by human pressure with a special focus on marine litter	45,45	0,00	22,73	31,82	0,00	0,00
	Water sustainability	30,00	0,00	35,00	25,00	10,00	0,00
	Marine litter and plastic pollution: toward a comprehensive and inclusive set of solutions	73,33	0,00	20,00	6,67	0,00	0,00
	Baltic Sea anatomy	22,22	0,00	11,11	55,56	11,11	0,00
Sustainable fishing and aquaculture	Sustainable aquaculture—what is at stake for the future?	24,11	2,84	13,48	49,65	9,93	0,00
	Will Italian fish products disappear from our dishes?	58,33	0,00	10,00	31,67	0,00	0,00
	Responsible Research and Innovation for Sustainable Fishery and Aquaculture: Opportunities and synergies between the fisheries and aquaculture sectors in Azores	44,64	3,57	7,14	35,71	8,93	0,00

(continued)

Table 10.2 (continued)

MML workshops		RRI dimension					
		Governance (%)	Ethics (%)	Public engagement (%)	Science education (%)	Open access (%)	Gender (%)
Renewable energy	Wave Energy in Ireland: Considering the future—perspectives on policy	12,99	16,88	25,97	19,48	22,08	2,60
	Wave energy in Ireland: Considering the future—an industry perspective	21,05	7,02	29,82	19,30	15,79	7,02
Marine changes caused by climate	To adapt and survive or to die. How to maintain and enhance the quality of the Black Sea NW coastal ecosystems under the pressures of climate change and human interventions	14,29	8,93	14,29	35,71	26,79	0,00
	Average	36,20	7,44	21,56	24,82	9,42	0,57

Concerning the relevance of RRI dimensions, Table 10.2 shows the percentages of actions per workshop and per RRI dimension.

The analysis of the MML workshops highlighted that Governance was the most frequently addressed dimension of actions proposed by the participating stakeholders with an average value of 36,20%. Actions involving Governance were connected to all marine challenges and were mainly suggested during the workshops on Pollution Caused by Human Land and Sea Pressures litter ("Marine litter and plastic pollution: toward a comprehensive and inclusive set of solutions" and "Marine litter: from the land to the sea and then back to us in our seafood?"), Sustainable Fisheries and Aquaculture ("Will Italian fish products disappear from our dishes?"), and Tourism and Coastal cities (Connecting Harbour and city—Strategy, collaboration and growth). Other predominant dimensions were Science Education and Public Engagement. Therefore, the first observation is that the most urgent need is for policymakers to elaborate norms and laws for improving the governance processes related to Pollution Caused by Human Land and Sea Pressures, Sustainable Fisheries and Aquaculture and Tourism and Coastal Cities. Moreover, Governance seems very relevant for all MML workshops related to Pollution Caused by Human Land and Sea Pressures", with a great expectation of concrete actions from all actors. In relation to "Fishing and Aquaculture", the most relevant dimensions are Governance and Science Education, showing how progress can be obtained combining good norms and regulations with the improvement of stakeholder and citizen knowledge. The challenge of Tourism and Coastal Cities appears as a very multifaceted topic. In fact, all the RRI dimensions (apart Gender) are addressed in a balanced way there. This requires a long lasting solid collaboration and active participation of all stakeholders in implementing the identified actions and defining solutions. The marine challenges of Marine Change Caused by Climate and Renewable Energy appear multifaceted too, but with the highest weight of Public Engagement, Science Education and Open Access. This highlights the important role that citizens, universities, schools, science centres, aquariums and enterprises can play in facing these challenges.

The technical subsystem (WKSP) of MARINA provides an open access to the detailed results of the Mobilisation and Mutual Learning debate (reports, working documents, scientific papers, videos, online chats, full-text analysis, workshop promotional material etc.) enabling to filter information by the RRI dimensions, Marine Issues, Societal Challenges and other aspects and to extend the knowledge sharing beyond the boundaries of the MARINA project.

10.4 Conclusion

This paper discussed the socio-technical approach of the MARINA project presenting the first results obtained involving 402 stakeholders in the 17 MML workshops across Europe and in the WKSP. The debate on the RRI dimensions and marine issues highlighted a need for Governance in relation to marine issues, but at the same time, the necessity to share responsibility (Public engagement) and the accessibility to knowledge (Science Education and Open access).

References

Coakes J, Coakes E (2005) Sociotechnical concepts applied to information systems, management information systems, vol 7. Blackwell Encyclopedia of Management, pp 281–286. Lehaney B, Clarke S, Coakes E, Jack G (2004) Sociotechnical systems and knowledge management, beyond knowledge management. Idea Group Publishing, Hershey, PA, pp 31–75

Dearden A, Rizvi H (2009) A deeply embedded sociotechnical strategy for designing ICT for development. J SocTechnol Knowl Dev 1(4):52–70

Ferri F, Grifoni P, Caschera MC, D'Ulizia A, Pratico C (2013) KRC: KnowInG crowdsourcing platform supporting creativity and innovation. Adv Inf Sci Serv Sci 5(16):1–15

Ferri F, Grifoni P, Caschera MC, D'Andrea A, D'Ulizia A, Guzzo T (2014) An ecosystemic environment for knowledge and services sharing on creative enterprises. In Proceedings of the 6th international conference on management of emergent digital ecosystems MEDES 2014, Buraidah, Al Qassim, Saudi Arabia

Grifoni P, Ferri F, D'Andrea A, Guzzo T, Praticò C (2014) SoN-KInG: a digital eco-system for innovation in professional and business domains. J Sys Inf Technol 16(1):77–92

Regulation (EU) No 1291/2013 of the European Parliament and of the Council of 11.12.2013 establishing horizon 2020-the framework programme for research and innovation (2014e2020) and repealing Decision No 1982/2006/EC. Off J Eur Union

Chapter 11
Humanitarian Laboratory for Syrian Crisis Transformation: A Project Proposal Towards Smoothing the Path for Achieving the Sustainable Development Goals (SDGs) in War-Torn Syria

Amira Buz Khallouf

Abstract The global response system to the Syrian crisis has failed to stop the war in Syria during the last six years. Tackling the humanitarian dimension of this catastrophe has become a huge global challenge. This position paper presents a project proposal for dealing with it via innovation and applying RRI framework. It is based on addressing it through multidisciplinary evidence-based research by providing in-depth, community-based, and a comprehensive analysis of the current contexts. It aims at identifying the various existing opportunities and gaps. Therefore, for the aim of achieving this end, we propose establishing a Humanitarian Laboratory for Syrian Crisis Transformation. It will be composed of six basic pillars: Economic Unit, Political Unit, Environmental Unit, Health Unit, IT Unit and Societal Integration and Recovery Unit. Each unit will include highly specialized humanitarian researchers, experts, inventors and scientists. This, in turn, will help in exploring all the potential ways for achieving SDGs agenda in Syria and guiding post-conflict reconstruction phase.

11.1 Introduction

The Syrian humanitarian crisis has been considered the greatest humanitarian tragedy of our times (Anon 2017). The global response system failed to stop the conflict and to tackle the humanitarian dimension of the catastrophe. As the crisis continues to displace more people, satisfying their basic needs has become a global challenge. Hence, applying (RRI) framework in the process of addressing the Syrian crisis has become an urgent need. It requires involving early a wide range of

A. Buz Khallouf (✉)
Syrian Social Innovators, Lisbon, Portugal
e-mail: sy.socialinnovators@gmail.com

F. Ferri et al., *Governance and Sustainability of Responsible Research and Innovation Processes*, SpringerBriefs in Research and Innovation Governance,
https://doi.org/10.1007/978-3-319-73105-6_11

actors and publics in all stages of practice, deliberation, and decision-making to maximize the quality of the collaboration work (Rri-tools.eu 2017).

This position paper presents a project proposal for approaching the Syrian crisis via innovation. It aims to address it through multidisciplinary evidence-based research by providing in-depth, community-based, and a comprehensive analysis of the current contexts plus identifying the various opportunities and gaps in its scenario. This, in turn, can smooth the path for achieving the SDGs within the Syrian communities outside and inside the country. For the aim of achieving this end, we propose establishing a humanitarian laboratory composed of six basic pillars: Economic Unit, Political Unit, Environmental Unit, Health Unit, IT Unit and Societal Integration and Recovery Unit. Each unit should consist of specialized humanitarian researchers, experts, inventors and scientists. Thus, finding practical solutions depends on inter- and intra-unit collaboration. Expanding the spectrum of the frameworks of these units will benefit the global refugee crisis in general and will help in exploring all the potential ways for achieving SDGs agenda in Syria and guiding post-conflict reconstruction phase. Undoubtedly, this laboratory will allow us to apply the anticipative and reflective mechanisms of RRI and explore the underlying assumptions and strengths that can better sharpen our vision while acting on addressing it. Hence, we can build participatory Research and Innovation (R&I) actions and provide inputs to influence policies.

11.2 Project Proposal of the Economic Unit: Constructing an Autonomic, Multi-governed and Sustainable Economic System for Refugees

Dealing with the global refugee crisis through the traditional U.N. approach has proven the failure of the global humanitarian response system. Managing it requires adopting new mechanisms and more comprehensive strategies (Khallouf 2016). Hence, in order to promote humanitarian crisis recovery on a multidimensional scale, the implementation of a Strengths-Based Approach to Refugee Crisis Response is of crucial importance at this stage. It revolves around considering refugee communities a huge source of human capital[1] (Econlib.org 2017), smoothing the path for investing in their latent capacities and qualities, and turning from focusing on the features of vulnerability into stressing the value of refugees' physical and intellectual potency. In practical terms, we need to identify strengths and weaknesses of both of refugees and host communities then to optimize the strengths of refugee communities in order to supplement the weaknesses of host

[1]Gray Becker used this term to refer to knowledge, skills, health, experiences, and values owned by the individual, it is a renewable source of creativity and innovativeness. The best investment in human capital can be done through education, training, health care, etc. Recruiting this stock of capacities in the appropriate fields yields economic values.

countries. Using strengths and weaknesses as tools of crisis transformation will turn the process of hosting refugees from a burden into a benefit to both sides.

This proposal does not imply allocating a certain geographic area somewhere on the planet for settling the global refugee communities, that would be governed by an independent political system as an essential stage prior to establishing an economic system. Instead, it aims at constructing an autonomic sustainable economy, multi governed by all refugee communities existing in different regions of the world in order to generate positive and multidimensional impacts and change the quality of life of refugees and host countries as well.

Nowadays, corporations play an influential role in people's life around the globe. A large proportion of the economies of the world is run by huge corporations whose market caps are larger than many countries' GDP. As a matter of example, the market value of Apple has become bigger than that of the entire Belgian market. Furthermore, many studies have emphasized that refugees can contribute to host countries' economic growth through boosting consumption in their markets. International credit-rating agency Standard and Poor's said that the refugees contribute to Turkey's economic growth and can raise Turkey's headline growth by 0.2–0.3% points per year (Daily Sabah 2017). However, instead of focusing a large attention only on the growing consumption rates which cause an increase in the economic growth of host countries, the importance of productive capacities of refugees and what they can add to the host countries, not only economically but also politically, environmentally and socially speaking, must be highlighted.

If refugees could establish their own economy and provide domestic, regional or global markets with a variety of goods and services that can foster effective competition or even build networks between substitutable and complementary products among all these markets, this process will promote consumer well-being, contribute to the growth of these economies and help integrate refugees socially within other communities. In very simple terms, we can picture the global refugee communities as a huge human capital of an immense global corporate group that is not functioning due to the absence of financial resources, organizational framework, regulations and work strategies. The biggest challenge is to find an appropriate answer to the key question of how to overcome the point of lacking the four previously mentioned components and to make use of the human capital effectively in order to make this "corporation" function.

For the aim of maximizing the possibilities of solving this question, The Brundtland Report of 1987 is turned to. It defines sustainability as "development that meets the needs of the present without compromising the ability of future generations to meet their own needs." This report mentioned that the three mutually dependent and interrelated areas of sustainability are economy, environment, and society (Development 2017).

Changes that happen in one area will definitely affect the other two areas, thus the functions of these circles overlap with each other. For example, non-renewable resources are usually used in the processes of achieving economic growth and that, in turn, affects the environment and the social well-being of our societies by producing harmful emissions and generating negative multidimensional impacts on people's lives

(Brundtland 1987; Elsevier and Higgins 2013). Thus, while reflecting on how to create a sustainable economic system for refugees, the other two interrelated cycles of this paradigm: society and the environment can bring the breakthrough at this stage. Both of them can play a major role in constructing and supplying the economic system with its essential components. The varied reservoirs of renewable natural resources that are available in host environments, diversity in cultural and social backgrounds among refugees and host societies besides to physical, intellectual and human resource endowments, all of these elements should be used as stepping stones on the road towards launching this economic system. Constructing this independent economic system for refugees requires the approval of the international community along with the host countries that can provide refugee communities with a wide range of facilities starting from regulatory procedures and the lands that have been occupied for several years by camps of refugees. They should be utilized now to host the huge conglomerates of this economic system. In addition, making a global call for international funds to be given as long and intermediate-term loans for new start-ups and entrepreneurship that have the potential to bring investments and money making projects and participate in the process of transforming these camps into smart, high-tech, industrial and eco-friendly cities. These types of cities can contribute to making advancements at several levels. Thus, they can stimulate movements of people and capital from different corners of the globe. Founding businesses based on clean industries, maintaining environmental integrity and human well-being, promoting tourism and social interactions, will bring about the radical solution to the global refugee crisis. Furthermore, this proposal calls for creating a shift in modelling our global economic systems by moving from the linear models of economies to circular ones. Indeed, emerging from this global humanitarian crisis requires building money making projects and giving loans over short, intermediate and long periods of time to refugees instead of leaning on donations for addressing refugees' basic needs. Shifting the frameworks of humanitarian contexts from living on charity to benefiting from huge investments in people's capacities, designed through systemized global plans, will radically revolutionize the harsh realities of humanitarian contexts.

11.3 Project Proposal of the Social Integration Subunit: Planning for Social Change in Syrian Refugee Communities Through Leadership Programs

The importance of leading change in Syrian refugee communities stems from the fact that it will better integrate the Syrians into host countries and help them contribute meaningfully to the local societies and economies.

How can a host society stimulate a planned change in a Syrian community without causing a radical damage in the genuine image of the Syrian society?

It can be done through preparing leadership development programs focused on building the leadership character from psychological, social and political perspectives and attracting leadership talents from both communities—namely Syrian and host communities. They should select several leaders who reflect the diversity of their communities. Hence, through these programs, each leader should acquire the appropriate skills to become an influential leader, capable of adapting to new and difficult situations and emerging from crises. They should learn about themselves and their societies: what their strengths and weaknesses are. Such programs will teach the Syrian leaders the key elements of success in leading the Syrian fragmented society. They will explore how to build creative strategies that can address the social and political damage and provoke positive and constructive change in their society without causing harmful impacts on its deep and genuine roots. Undoubtedly, cooperation between the leaders of both communities will lead to better integrate the Syrians into host communities and to shorten the social distance between both communities. Part of responsibilities of the Syrian leaders lies on inspiring the other Syrians, focusing on developing their skills and talents and preparing them to be agents of change, who will draw the picture of the future of Syria in a harmonized manner. Besides, working on strengthening community ties is an essential task that can foster strong links with the population of the home land on one hand and channel the Syrian potentials into service of host countries, on the other hand. At the core of leaders' mission lies directing the concerns of the Syrian minds towards setting plans of Post-Conflict Reconstruction in Syria. Given the fact that the impact of brain-drain on the future of Syria will be immense, this is why promoting the process of brain gain and encouraging the Syrians to work on building their country by using the knowledge and experiences gained abroad is crucial for the period of reconstructing the country.

11.4 Project Proposal of the Social Recovery Subunit: Establishing the Hypothesis of Social Genetics Towards Making Sense of the Global Refugee Crisis

Genetics is the study of heredity, a biological process in which parents pass certain genes onto their offspring. Behavioral Genetics has revolutionized our understanding of the origins of individual differences in behavioral traits. Undoubtedly, looking at the interplay between genetics and environment is of vital importance in the process of building the theory of social genetics. Thus, the interaction that happens between our genes and environments influences our traits and behaviors. The environment is everything that affects the individual except his genes (Internet Archive 2017). Besides, the environment has social, cultural and physical forces that affect largely the growth and development of the individual.

Let's suppose that we have another kind of heredity that occurs through social processes called social heredity. Thus, social genetics acts on socially constructed concepts such as laws, religions, cultures, traditions and customs, identity, beliefs, mother tongue, work of art, philosophical views and any element of political, economic and social systems. Let's suppose that we have social genomes that contain social genes. Certainly, they do not exist in a physical sense. However, these social genes should play an influential role in shaping our behaviors and patterns of social interactions. Thus, they contribute to forming the diverse types of social structures, social processes and problems, and their interrelationships.

Indeed, the genes of social dynamic change are the first cluster of social genes that we will identify. Having them provides individuals with mechanisms of transformational change that can impact any life aspect. They act on modifying the mindsets of the individuals and that, in turn, will lead to modifying their behaviors and attitudes. They can cause the shift from fixed into growth mindset as people's attitudes fall into one of two categories: a fixed or growth mindset according to Dweck (2008).[2] These genes can stimulate growth at several levels, more importantly: social, economic and political. Thus, different social constructs will grow into multiple modalities as an immediate result of the growth of the individuals' mindsets. Hence, expressions of these genes give rise to polymorphisms of humans' mindsets and consequently to polymorphisms of socially constructed concepts.

As a matter of example, this will allow the same religion to develop and grow into multi-diversified modalities instead of having it fossilized in a fixed and narrow template. Thus, several religious approaches will arise to complete each other instead of competing with each other.

Let's link the hypothesis of Social Genetics to the current migratory phenomena: refugees. Refugees belong to a wide range of social and cultural backgrounds. They flee mainly due to political conflicts and economic and social challenges, which are the result of incongruent goals and attitudes within their own societies. This destructive incongruence happens mainly due to dysfunctions of the social genes of individuals of these societies. Therefore, turning social gens off and on, removing and implanting others, and identifying and designing constructive mechanisms that govern the nature of interactions among the social genes and their polymorphisms is intended to be a matter of investigation.

However, resolving inter- and intra-social concepts' conflicts and putting them into a state of peace, coexistence, harmony and cooperation requires primarily identifying:

- Pools of social genes.
- Clusters of social genes that provide individuals with effective mechanisms for tackling destructive incongruence and boosting permanent growth at several levels in life.
- Clusters of social genes that hinder growth and give rise to huge

[2]Carol Dweck's theory of mindset has differentiated the fixed from the growth mindset. It explains how our belief systems about our own capacities and potential are able to fuel our behaviors and actions and contribute to our success in life. According to her research, "success is all about how you deal with failure". People with a growth mindset welcome challenges and see them as opportunities while those with a fixed mindset feel hopeless and overwhelmed when their challenges go beyond their control.

- problems such as social division, social exclusion, discrimination, racism, inequity, violence, conflicts and wars.
- Mechanisms of social genetic manipulation that cause or even affect
- change in the social genetic makeup of a social organism.

In sum, social genes act at both micro (small-scale social processes such as interactions between individuals in small groups) and macro (large-scale social processes) levels. They contribute largely to the architecture of the mental framework of each individual and establishing the complex framework of each society. Dysfunctions of societies arise from maintaining the dysfunctional social genes over a period of time by transmitting them from one person to another and across generations. As a result, the dysfunctional social genes are the latent hands behind inter- and intra-groups division, conflicts and wars.

11.4.1 Social Gene Manipulation

In order to identify the mechanisms behind changes in societies, building on the theory of Social Physics can provide us with a comprehensive understanding of the critical role that ideas and information flow play in the process of forming societies and their structures. It describes how the social phenomenon is shaped by the social networks and all patterns of transactions between individuals. Not only money and goods people exchange but also information. That all translates into changes in behaviors and attitudes as people learn from each other and influence each other's opinions (Pentland 2015). In other words, Social Physics identifies mathematical connections between idea flow from one person to another on the one hand and people's behaviors and actions on the other. Thus, spreading and integrating novel ideas drives innovation and behavior change (Pentland 2015).

Social Genetics is a quantitative and qualitative social science. It attributes the environmental forces that take part in forming human attitudes and interactions within their networks plus the underlying mechanisms that drive change in individual and collective actions to hereditary elements running in each established social organism. Similarly to Social Physics, Social Genetics recognizes the critical role of information flow in shaping elements of social systems. Social Genetics considers that the mode of information flow and the characteristics of information networks, namely: the relationships between the individuals, the nature of ideas exchanged between them and the rates of information flow are instrumental in forming the nature of any element in societal structures.

11.4.2 What Is Social Genome?

Social Genome is the latent reserve of all societal codes. It is the complete set of hereditary social instructions. It contains all the information contributed to building a social organism. Besides, it includes the underlying predictors to societal crises, failure, success, weaknesses, drawbacks and strengths. The instructions of social genome are formed mainly by the nature of the relationships between the individuals of the social organism. It lies behind the diverse modes of information flow within the networks of social organism. Social genes code for individual and collective behaviors and formation of structures and systems of social organisms. Social genes take a large part in fueling the different types of behaviors such as good and evil deeds, racism, discrimination, persecution, inequity, corruption, justice, injustice, empathy, etc. However, each individual has a unique social genome. The human social genome is made up of codes of models of information flow. They are parallel to the biological bases of DNA. However, the social genome sizes differ from a social organism to another. Social organism's inheritance depends on the structures and functions of the modes of information flow. According to Social Genetics, bringing innovative ideas requires manipulating the nature of several ideas as well as the networks through which these ideas pass. Hence, modifying structures, functions and rates of information flow are central to any change. That, in turn, will facilitate ideas connections, combinations and growth. While information is flowing in several directions and channels, novel ideas are continuously generated.

11.4.3 How Can the Hypothesis of Social Genetics Address the Global Migration Crisis?

In principle, it can be done through creating national and international banks of human social genomes and setting parameters for human genomes interactions to predict direct and indirect consequences and design suitable policies for phenomenon like migration and massive human influx due to humanitarian crises. In other words, setting parameters identifying the underlying role of migration on the levels of growth and development of societies through studying channels of information flow can shift the art of dealing with migratory phenomenon radically.

Besides, constructing vivid models of genetically modified social organisms via applying mechanisms of social gene engineering that have the potential to bring about solutions to pressing challenges and remedies to societies, in particular, those overwhelmed by conflicts will represent an important turning point.

Guiding Tips towards Modifying Mindsets of Refugees:

- Re-linking refugees strongly to their origin communities and, most importantly, their social constructs instead of focusing the efforts solely on integrating them into host countries:

  This process will pave the way for tackling the damage of the social fabric of refugees' communities and will consequently lead to reconciling and reuniting the populations in the future.

- Building a Construct of Refugee Beliefs:
  If the notion of "the affected population is decisive tools in transforming conflicting societies" has become a belief, refugees will no longer perceive themselves as vulnerable, rather agents of change who will perceive the challenges of their realities as opportunities.
- Breaking refugees' social concepts up into several components and reconstructing them again:
  Constructing new models out of the old ones, such as creating new religious approaches based on embracing the conflicting religious ideologies in unified modalities represent the primary stage in this process. Putting the new models into practice and directing them into channels of permanent modification while transmitting them among individuals and across generations will make refugees see diversity as a motor force for prosperity and a sign of strength of their societies instead of perceiving it as a catalyst for division and conflicts.
- Refugee Brain Reconstruction:
  In principle, mental imagery approach is an essential step for guiding refugees into forming a positive mental picture about their realities, especially, those who fled due to wars and extreme violence and suffered traumatic experiences. Collective re-visualization of their crises and current contexts will help them invest in building networks of collaborative work among each other for the aim of transforming their harsh realities into the mentally visualized image. As a result, solutions will move smoothly from ideas level into action level. That, in turn, will pave the path for achieving the SDGs in their communities.

11.5 Conclusions

In sum, the mission of this laboratory is based on channeling collaborative efforts into exploring new paths for addressing the Syrian humanitarian catastrophe in particular plus the global refugee crisis in general by investing in the valuable physical and intellectual potentials of refugees. This humanitarian laboratory aims at bringing novel solutions by digging deep in refugees' contexts through applying the RRI process dimensions. For instance, through implementing the RRI responsive and adaptive dimensions, we will be able to modify modes of thought, actions and organizational structures, in response to changing circumstances and perspectives. It will enable us to uncover existing opportunities that can complement the unfinished work of the international community after failing dramatically to solve the most pressing challenge of our time: refugee crisis.

References

Anon (2017) https://ec.europa.eu/echo/files/aid/countries/factsheets/syria_en.pdf. Accessed 5 Jun 2017

Brundtland G (1987) Our common future, 1st edn. Oxford University Press, Oxford

Daily Sabah (2017) Refugees contribute to Turkish economy, S&P says. https://www.dailysabah.com/economy/2016/05/10/refugees-contribute-to-turkish-economy-sp-says. Accessed 6 Jun 2017

Development W (2017) Our common future. Chap 2 Towards Sustainable Development-A/42/427 Annex, Chap 2-UN documents: gathering a body of global agreements. Un-documents.net. http://www.un-documents.net/ocf-02.htm. Accessed 6 Jun 2017

Dweck C (2008) Mindset, 1st edn. Ballantine Books, New York

Econlib.org. (2017) Human capital: the concise encyclopedia of economics. Library of Economics and Liberty. http://www.econlib.org/library/Enc/HumanCapital.html. Accessed 6 Jun 2017

Elsevier, Higgins KL (2013) Economic growth and sustainability–are they mutually exclusive? https://www.elsevier.com/connect/economic-growth-and-sustainability-are-they-mutually-exclusive. Accessed 6 Jun 2017

Internet Archive (2017) Foundations of psychology: Boring, Edwin Garrigues: Free Download & Streaming: Internet Archive. https://archive.org/details/foundationsofpsy00bori. Accessed 6 Jun 2017

Khallouf A (2016) Recommendations for transforming the Syrian humanitarian crisis. Procedia Eng 159:272–274

Pentland A (2015) Social physics, 1st edn. Penguin Books, New York, NY

Rri-tools.eu. (2017) About RRI-RRI tools. https://www.rri-tools.eu/about-rri. Accessed 5 Sep 2017

Chapter 12
Responsible Innovation: A New Approach to Address the Theoretical Gaps for Innovating in Emerging E-Mobility Sector

Shilpa Mishra and Rajbeer Singh

Abstract Sustainability along with being an 'after-work' of technology has become more about solving the problems and issues which pave way for a robust approach towards social and technological development. The emerging technologies are promising to fulfil the need of society alongside they have also posed tremendous challenges in the form of new emerging risks, unforeseen impacts, governance, ensuring accountability for results of adoption and sustainability. The existing approaches towards innovation such as NIS, SIS, TIS, sustainable innovation, and transitions are unable to address issues of long-term future impacts and issues of accountability, sustainability, and risks from innovators as well as consumers point of view. The new approach of responsible innovation claims to address the above-mentioned issues through bridging the existing gaps in the theories of innovation. Therefore, this paper focuses on two research questions, how the deliberation and participation enabled adoption of new technology? What are the different values embedded by the e-mobility (e-rickshaw) innovation that shapes it as a responsible innovation? To address these questions a case study "adoption of e-rickshaws in Delhi" has been selected. It was observed that RI approach enables us to understand the process of innovation for emerging technology in mobility area i.e. adoption of e-rickshaw in Delhi. The two dimensions of responsible innovation enable us to understand that how the process of innovation unfolded and how values have been embedded in innovation and the understanding can be used for innovating in newly emerging areas of technology.

S. Mishra (✉) · R. Singh (✉)
School of Social Sciences, Centre For Studies in Science Policy, Jawaharlal Nehru University, New Delhi, India
e-mail: shilpa67_sse@jnu.ac.in

R. Singh
e-mail: rajbasera@gmail.com

R. Singh
School of Innovation Sciences, Department of Industrial Engineering & Innovation Sciences, Eindhoven University of Technology, Eindhoven, The Netherlands

F. Ferri et al., *Governance and Sustainability of Responsible Research and Innovation Processes*, SpringerBriefs in Research and Innovation Governance,
https://doi.org/10.1007/978-3-319-73105-6_12

12.1 Introduction

Society and technology together make socio-technical system. The shift from one socio-technical system to another raises the issue of sustainability, responsibility, accountability towards adverse impacts of emerging socio-technical system. Presently, the socio-technical system can be identified with developments as the Internet of things (IOT), big data, smart cities, electric mobility, smart grids, smart metering, social media, nanotechnology and renewable energy technologies. Not only sustainable outcome of these socio-technical systems are desirable, but their sustainable implementations are also required. Apart from the sustainability, the issue of risk, ownership, accountability, responsibility are also present in a new socio-technical system. These innovations are created through collaborations and partnerships with many individuals, firms, and universities. The emerging technologies require new approaches to governance (Guston 2013). Innovation is not created within boundaries of one company or organization or even countries. It emerges from multiple locations, organization, companies, universities, and individuals embedded in a network innovation and creativity. This situation has scattered the risk and accountability. It is hard to identify accountability as an entity for any negative impacts or consequences due to its adoption (Setiawan and Singh 2015). As innovation process is becoming more and more open (Chesbrough 2006), it is leading to less control of the innovator. The present context leads to crossing all the boundaries of rules and standards in practice, within a country or society. It means the existing rules and standards cannot govern innovations e.g. civilian use of drones and use of e-rickshaws in India (before the year 2013). Therefore to understand the process of innovation and its adoption, one need to look through precise theoretical understanding and approaches, and the exploration of different approaches would help in developing such an understanding.

12.2 Responsible Innovation

First, we should discuss sustainability. After some international commitments like Rio 1992, it is evident now, sustainable development meaning is to achieve long-term goals is technologies infrastructure, lifestyles, and institution (Rennings 2000). The eco-innovation and sustainable innovation are a process of developing a new approach, products, and process to reduce the environmental burden (Renning 2000). The sustainable innovation concept has helped policymakers to a large extent, to achieve sustainability goals. But theoretical frameworks like sustainable and eco-innovation do not address an issue of accountability and responsibility. The notion of responsibility in common public understanding refers as the capability of fulfilling the obligation. Duty performs by an individual and his/her obligation towards some work called as responsibility (Ubois 2010). Being accountable refers to the sense of being called on to answer for the action or discharge of duty.

Accountability ensures that there is answerability for some undesirable action. Adoption of innovation changes existing responsibilities and accountability in the social system, in this situation sharing and allocating of new responsibility and accountability among stakeholders is needed, sharing of responsibility shifts the individual blame approach to system blame (Geels et al. 2004). Individual blame is a tendency to hold and make someone individually responsible for his or her action, on the other hand, system blame all elements of innovation with a system framework are responsible (Rogers 1983). The responsibility of outcomes derived from innovation is not related to a single actor of innovation (Von Schomberg 2013). In this situation, a theoretical framework is required that can allocate responsibility and accountability through a participatory process. It is important to decide responsibility in starting of adoption of innovation; this helps in anticipation of impacts (negative) and its prevention through technological modification and institutional arrangements (rules and regulations). We should care both transformative advantages and adverse impacts of them (Stiloge et al. 2013). Responsible research and innovation should be understood as a strategy of stakeholders to become mutually responsive to each other, anticipating research and outcomes aimed at the "grand challenges" of our time for which they share responsibility (Von Schomberg 2013). Responsible innovation like umbrella covers technology assessment, anticipation, and governance. In the innovation process, large numbers of actors are involved, so the outcome of innovation is seldom a result of the single irresponsible actor. There is no unison on the concept of irresponsible innovation. Still, it is not very clear that which type of innovations come under irresponsible innovation. Some authors like Schomberg has categorized the four types of innovations under irresponsible innovation: Innovation that is the result of technology push, Innovation that neglects fundamental ethical principles, Innovation that is introduced through policy pull, and Innovation that is introduced in the social system without anticipation of consequences of innovation (Von Schomberg 2013). In the case of cross culture transfer of technology, if innovation is not sensitive to culture-specific value, then it is not a responsible innovation (Setiawan and Singh 2015). The erstwhile definitions of innovation have something in common; innovation encompasses (potential) consequences upon socio-economic and ecological systems, which are against the values of those who are affected and concerned with innovation. "Responsible innovation is transparent and interactive process involving societal actors and innovators in a mutual and responsive way with a view to (ethical) acceptability, sustainability, and societal desirability of the innovation process and its products" in order to allow a proper embedding of scientific and technological advances in our society (Von Schomberg 2013). Richard Owen has defined responsible innovation such as "it is a process which undertaken towards, socially desirable and socially acceptable ends, with the connotation of trust and integrity" (Owen et al. 2013). From preceding two definitions, it reflects that aim of responsible innovation is to provide such type of innovation that should be socially acceptable and desirable with the ultimate vision of sustainability. Singh and Otto (2012) defined responsible innovation as, "being care or ensuring care for certain values for social, economic and environmental sustainability by engaging participation, reflexivity, deliberation, responsiveness and participation for bringing up any change in any idea, product, process, method, way of doing business, technology in order to bring them into a specific market or use them in society" (Singh and Otto 2012).

12.3 Case Study: Description

Since ancient times, travelling has been considered necessary for knowledge, economic growth, and social progress. But, with the passage of time mobility has become a challenge for cities, and the concept of sustainable mobility has got immense importance. The transport system is a key driver in attempts to increase the degree of sustainability in cities. Under goal eleven of the sustainable development, goal two is to provide access to safe, affordable, accessible and sustainable transport system for all, using road safety and by expanding on public transport (United Nation 2015). Sustainable mobility is understood as the means to contribute to social and economic welfare, without damaging the environment for depleting environmental resources. The world business council for sustainable development has defined sustainable mobility as the ability to meet the needs of society to move freely, gain access, communicate, trade, and establish a relationship without sacrificing other essential human or ecological values in the future (Nykvist and Whitmarsh 2008). India and other Asian countries are experimenting with new transport technologies and mobility means to achieve the sustainability goals. Larger jumps, in environmental efficiency, can be achieved only by radical technologies (Geels et al. 2004). The innovation of E-rickshaw is a step in that direction and it also a technological revolution. Adoption of e-rickshaw is a transition from a fuel-based technological regime to electricity based technological regime. E-rickshaws were adopted in 2010, and since then their number has been increased (Shanawaz Khan Versus Municipal Corporation of Delhi and Others 2014). But, the ban on E-rickshaws had shattered the hope of E-rickshaw drivers (who are already adopted it) and cycle rickshaw pullers (who want to adopt it), so they started protest rally against the ban. Meanwhile some studies focus on consumer behaviour for adoption of this environmental friendly innovation, role of national government for promotion of this disruptive innovation for employment, and actors negotiations to achieve acceptability of this innovation (Dangi et al. 2013; Chandran and Brahmachari 2015; Ghosh et al. 2016). Hence soon this technology (E-rickshaw) has witnessed interesting interplay between state institutions, political actors and technology developers (Ghosh et al. 2016). The debate has started in the judiciary, parliament, and media. Now questions emerged, how these debates identified the loopholes of e-rickshaw based transport medium? How this innovation is being adapted in Indian transportation system? There are various key elements to promote the public acceptability of sustainable mobility, and adaptability is one key factor. Adaptability provides the platform for transparent decision-making, leadership, supported by analysis and monitoring to check the effectiveness of policy action (Banister 2008).

12.3.1 Analysis of Case

This study is an exploratory type. The data has been collected through document analysis, semi-structured interview and focused group discussion. The innovation development process (from identification of the problem to the adoption of innovation) is consisting of a vast number of actors or stakeholders from individuals to corporate and states (Lee and Petts 2013). The term stakeholders refer to "persons, group or organization that must be somehow taken in account by leaders, managers, and frontline staff" (Bryson 2004). A stakeholder is any actor that can affect or can be affected by, a decision or action. Responsible innovation framework emphasizes the involvement of actors in each step of innovation that makes innovation acceptable. The two dimensions of responsible innovation; deliberation and participation have been traced through discussed methodology. There are numbers of evidence in the past, when technological revolutions happened, there had been debates, dialogue and at times near warfare about them (Sykes and Macnaghten 2013). And similar situation has been created by adoption of e-rickshaw innovation in India. The debates and critiques that explore the different aspects of innovation, and give consideration of all negative and adverse impacts along with positive to make innovation acceptable and workable come under the deliberation dimension of responsible innovation (Singh and Otto 2012). Participation means taking part into or being part of the certain process; it indicates the involvement of stakeholder of innovation and the desired innovation can result in undesired and unacceptable outcomes without ensuring participation at different stages of innovation process (Singh and Otto 2012). In this case Deliberation was happened: through committee, collaboration and work projects, judicial hearing, and focused group discussions in media and Parliament debate. The participant were ministry of road transport and highway, Ministry Of Heavy Industries and Public Enterprises, Ministry Of Environment, Forest and Climate Change, Delhi Transport Department, Delhi Municipal Corporation, Delhi Police, High court of Delhi, Supreme court of India, Non Government Organisation (NGO)—Centre for Science and Environment, International Centre For Automotive Technology (ICAT), The Energy and Research Institute (TERI), Manufacturer, Charging Service Provider, E-rickshaw Drivers and Public. The standards, certifications and accreditations are a tools a new form of governance (Von Schomberg 2013). New Rules, standards, were formed through the deliberation and participation process. In this case values like safety, protection, reliability, comfortability, competitiveness, profitability, affordability, employment, non-polluting, zero-emission are considered as the universal values. The social equity, gender equality, quality of life, standard of living, and ownership are highly important values for the developing countries like India.

12.4 Conclusion

The main objective of this paper is to find out the evidence of responsible innovation theory in developing countries. In context of developed nation this theory is well established and tested. But developing countries are trying to find out evidence of theory. Deliberation and participation dimensions of theory have been tested, and values have been identified by using discussed two dimensions. So, this paper can be a good example to understand responsible innovation in developing countries perspective. The dimensions like deliberation and participation can be deploy any other study in developing countries. The values that identified for sustainability (social, economic, environment) are suited to developing countries. This paves the way for future research for responsible innovation theory.

References

Banister D (2008) The sustainable mobility paradigm. Transp Policy 15(2):73–80

Bryson JM (2004) What to do when stakeholders matter: stakeholder identification and analysis techniques. Public Manag Rev 6(1):21–53

Chandran N, Brahmachari SK (2015) Technology, knowledge and markets: connecting the dots—electric rickshaw in India as a case study. J Frugal Innov 1(1):3

Chesbrough HW (2006) Open innovation: the new imperative for creating and profiting from technology. Harvard Business Press

Dangi H, Laheri VK, Vohra (2013) A consumer behavior towards awareness and adoption of electric—rickshaws: a case study of Delhi

Geels FW, Elzen B, Green K (2004) General introduction: system innovation and transitions to sustainability. In: System innovation and the transition to sustainability, pp 1–16

Ghosh D, Sengers F, Wieczorek AJ, Ghosh B, Roy J, Raven R (2016) Urban mobility experiments in India and Thailand, chap 9. In: The experimental city, p 122

Guston DH (2013) "Daddy, can i have a puddle gator?": creativity, anticipation, and responsible innovation. In: Responsible innovation: managing the responsible emergence of science and innovation in society, pp 109–118

Lee RG, Petts J (2013) Adaptive governance for responsible innovation. In: Responsible innovation: managing the responsible emergence of science and innovation in society, pp 143–164

Nykvist B, Whitmarsh L (2008) A multi-level analysis of sustainable mobility transitions: Niche development in the UK and Sweden. Technological forecasting and social change, 75(9), 1373–1387

Owen R, Stilgoe J, Macnaghten P, Gorman M, Fisher E, Guston D (2013) A framework for responsible innovation. In: Responsible innovation: managing the responsible emergence of science and innovation in society, pp 27–50

Rennings K (2000) Redefining innovation—eco-innovation research and the contribution from ecological economics. Ecol Econ 32(2):319–332 Rogers EM (2010) Diffusion of innovations: Simon and Schuster

Rogers E. M. (1983) Diffusion of innovation: diffusion of innovation, The Free Press, New York, USA

Setiawan AD, Singh R (2015) Responsible innovation in practice: the adoption of solar PV in telecom towers in Indonesia, vol 2. In: Responsible innovation. Springer, pp 225–243

Singh R, Otto K (2012) Understanding responsible innovation developing countries perspective. In: 2nd conference on responsible innovation, The Hague, Netherland

Stilgoe J, Owen R, Macnaghten P (2013) Developing a framework for responsible innovation. Res Policy 42(9):1568–1580

Sykes K, Macnaghten P (2013) Responsible innovation–opening up dialog and debate. In: Responsible innovation: managing the responsible emergence of science and innovation in society, pp 85–107

Ubois J (2010) Responsible innovation/sustainable innovation. Paper presented at the annual meeting of the American anthropological association

United Nations (2015) Transforming our world: the 2030 agenda for sustainable development, Resolution adopted by the General Assembly on 25 September 2015, UNGA seventh session

Von Schomberg R (2013) A vision of responsible research and innovation. In: Responsible innovation: managing the responsible emergence of science and innovation in society, pp 51–74

Erratum to: Governance and Sustainability of Responsible Research and Innovation Processes

Fernando Ferri, Ned Dwyer, Saša Raicevich, Patrizia Grifoni, Husne Altiok, Hans Thor Andersen, Yiannis Laouris and Cecilia Silvestri

Erratum to:
F. Ferri et al., *Governance and Sustainability of Responsible Research and Innovation Processes*, SpringerBriefs in Research and Innovation Governance, https://doi.org/10.1007/978-3-319-73105-6

Chapters 1 and 3 have now been published open access under a CC BY 4.0 license. The original book has been updated with the changes.

The updated online version of the book can be found at
https://doi.org/10.1007/978-3-319-73105-6
https://doi.org/10.1007/978-3-319-73105-6_1
https://doi.org/10.1007/978-3-319-73105-6_3

F. Ferri et al., *Governance and Sustainability of Responsible Research and Innovation Processes*, SpringerBriefs in Research and Innovation Governance,
https://doi.org/10.1007/978-3-319-73105-6_13

MIX
Papier aus verantwortungsvollen Quellen
Paper from responsible sources
FSC® C105338

If you have any concerns about our products,
you can contact us on
ProductSafety@springernature.com

In case Publisher is established outside the EU,
the EU authorized representative is:
Springer Nature Customer Service Center GmbH
Europaplatz 3, 69115 Heidelberg, Germany

Printed by Libri Plureos GmbH
in Hamburg, Germany